Palgrave Studies in Impact Finance

Palgrave Studies in Green Finance

Series Editors
Helen Chiappini, Dipartimento di Economia Aziendale, D'Annunzio
University of Chieti–Pescar, Pescara, Italy
Mario La Torre, Facolta di Economia, Universita La Sapienza, Rome,
Italy

This subseries explores studies on green finance, specifically with regards to green investments, business models, roles of different actors in the green finance market, new regulatory and disclosure trends, green assets risks and performance alongside emerging areas including climate risk and green fintech with both theoretical and empirical approaches. Palgrave Studies in Green Finance represents the first international series dedicated to such a topic, meeting the growing interest of scholars, policy makers, regulators, and students in green finance.

Andi Duqi

Banking Institutions and Natural Disasters

Recovery, Resilience and Growth in the Face
of Climate Change

Andi Duqi
Department of Management
Alma Mater Studiorum—Università
di Bologna
Bologna, Italy

ISSN 2662-5105   ISSN 2662-5113 (electronic)
Palgrave Studies in Impact Finance
ISSN 2662-7388   ISSN 2662-7396 (electronic)
Palgrave Studies in Green Finance
ISBN 978-3-031-36370-2   ISBN 978-3-031-36371-9 (eBook)
https://doi.org/10.1007/978-3-031-36371-9

PREFACE

Humanity has tried to cope with the short- and long-term economic consequences of natural disasters for centuries. However, the severity of these events is constantly increasing in magnitude, due to the alteration of the earth's climate and the environment. They pose a serious threat to the lives of millions of people especially in less developed countries. At the same time, the international community has acknowledged that weather hazards' impact and countries' resilience to them is affected by various socio-economic factors, such as the well-functioning of financial institutions, especially commercial banks.

In this book, we aim to look more closely at how natural disasters impact bank activity, and how banks can support economic recovery after a natural disaster. The importance of banks in this context is underscored by increasing regulatory attention on their role in fostering a sustainable future, but also on the risks that climate change poses to bank stability.

Banks will inevitably adapt their strategies to address concerns arising from climate change. In doing this, they must partner with international cooperation institutions, other financial intermediaries, and local governments, so that the green transition ensures an inclusive growth for all, especially the most vulnerable parts of our societies.

Andi Duqi
Via Capo di Lucca
Bologna, Italy

ACKNOWLEDGMENTS

I would like to express my sincere gratitude to many people who helped make this book possible. First and foremost, I thank my family for their unwavering support and encouragement throughout the writing process.

I am also grateful to my colleagues at the Department of Management, Alma Mater Studiorum—Università di Bologna, and other Universities, with whom I discussed the topics presented in this book. Special thanks to Giuseppe Torluccio, Stefano Cenni, Giovanni Cardillo, Helen Chiappini, Danny McGowan, Aziz Jaafar, Enrico Onali, Chrysovalantis Vasilakis, Philip Molyneux, Dmitri Vinogradov, Salvatore Perdichizzi, Hussein Al Tamimi, Panagiotis Zervopoulos.

I am also indebted to AIFIRM, the Italian Association of Financial Industry Risk Managers, and AIFIRM members with whom I cooperated in finalizing the Position Paper 39/2022 on Bank Climate Stress Tests.

In addition, I would like to thank the Publisher, Palgrave Macmillan-Springer Nature, the Executive Editor, Tula Weiss, two anonymous referees, and other people in Palgrave-Springer who helped me craft the book. They provided invaluable insights that improved the overall quality of the book.

CONTENTS

LIST OF FIGURES

LIST OF TABLES

CHAPTER 1

Overview

Abstract Natural disasters have long-lasting socio-economic impacts on populations and countries impacted by them, especially in emerging economies. Weather-related catastrophes are becoming more prevalent as the planet heats up. These events affect the financial sector through several channels, such as the collateral value of bank loans or borrowers' ability to pay off debt. At the same time, the financial sector plays an essential role in the post-disaster economic recovery through reconstruction lending. The link between natural disasters and bank activity is, therefore, complex. This book aims to shed light on the importance of financial institutions in promoting growth after extreme weather events.

Keywords Natural disasters · Climate change · Physical risks

The natural environment is transforming; this alteration is influenced by how humans interact with nature, and how they are expected to transform their attitudes toward a changing natural environment. Weather-related catastrophes are becoming more prevalent, as the planet heats up. The intensity of some of them such as heat waves, tropical cyclones, and wildfires is amplifying considerably. A report by Barclays shows that the number of extreme weather events has augmented fivefold in the period

© The Author(s), under exclusive license to Springer Nature
Switzerland AG 2023
A. Duqi, *Banking Institutions and Natural Disasters*,
Palgrave Studies in Impact Finance,
https://doi.org/10.1007/978-3-031-36371-9_1

1970–2022, albeit diminishing slightly between the 2000s and the 2010s. Furthermore, extreme weather events have grown more intense and capricious with respect to timing and location, and their cost has soared almost eight times globally, inflation-adjusted, since the 1970s. This equates to cost per event soaring almost 77%, inflation-adjusted, over the past five decades.

Natural disasters have long-lasting socio-economic impacts on populations and countries that are impacted by them, especially in emerging economies. For instance, extreme temperature and humidity episodes, flooding and hurricanes, droughts and wildfires push populations to migrate, thereby destabilizing entire regions politically. These systemic social and economic changes might result in failed states and other breakdowns in countries in the most vulnerable areas. At the same time, companies may be exposed to drought, wildfire, floods, hurricanes, typhoons, or tornado risks depending on their activities. Studies have demonstrated, for example, how food production is affected by drought risk in many parts of the world, and how food companies' financial performance is increasingly impacted by these episodes.

Extreme weather events can severely affect and quickly destabilize financial markets and institutions. The financial sector is exposed to physical risks[1] through several channels. First, they affect the valuation of real estate assets, the main collateral of bank loans. Second, physical risks could hamper the borrowers' ability to pay off debt when they operate in sectors that could be particularly affected by disasters, such as agriculture or tourism.[2] Hence, the destruction and economic disruptions caused by hurricanes, wildfires, and other natural disasters may spill over to banks, particularly small, local institutions squared in the "eye" of the storm. If loan losses spike or customers move away over the longer run, bank solvency could be threatened. Indeed, the banking panic of 1907 was triggered by the earthquake and fire that ravaged San Francisco in 1906 (Odell and Weidenmier 2004).[3]

The financial sector plays an essential role in the post-disaster economic recovery. Banks provide emergency loans to businesses and households to

---

[1] Physical risks arise from climatic events, such as windstorms, floods, or tornados. They can be both acute or chronical. Natural disasters are classified as acute physical risks. This study uses "acute physical risks" and "natural disasters" terms interchangeably.

[2] Bolton et al. (2021).

[3] Blickle et al. (2022).

restore their damaged properties. They also channel government subsidies or disaster loans to the population. Therefore, banks' profitability could benefit from increased lending in the aftermath of a disaster. However, when banks suffer physical or capital damage due to catastrophic natural events, they may not be able to provide sufficient funds at the same interest rates as before. Borrowers' financial constraints could exacerbate when they need to invest for reconstruction (Hosono et al., 2016).

The link between natural disasters and bank activity is therefore complex. Our knowledge of how the banking sector is affected by extreme weather hazards and whether and how they contribute to the economic recovery after the disaster still needs to improve. Therefore, this book aims to shed light on the importance of the financial institutions in promoting growth after extreme weather events. Financial institutions are uniquely positioned to have a broad and sizeable impact on the challenges that natural disasters pose to humanity owing to their critical role in society. They serve as fiduciary stewards, allocators, and distributors of capital across the global economy and directly impact the ability of people to manage their money and build wealth.

At the same time, regulators and policymakers strongly emphasize and enforce the measurement and compliance of banks with risks related to climate change. Banks will be required to assess their exposure to climate risks correctly and contribute to channeling private investment toward climate-neutral and climate-resilient investment opportunities. In this context, banks could modify their strategies toward lending to disaster-prone areas, which could have serious consequences for the economic activity in entire world regions. Therefore, a careful understanding of bank behavior in the face of these risks and regulatory pressure is fundamental for evaluating post-disaster recovery, especially of less developed countries.

Banks are not the only relevant actor in restoring economic activities after natural disasters strikes. The insurance industry plays a vital role in dealing with the rising threat of extreme weather, appropriately pricing risks and losses from extreme weather events, and protecting against them. Evidence shows that insurance earnings are directly affected by extreme weather. While nominal catastrophe pricing has improved since 2017, risk-adjusted prices have little changed—indicating that insurers are not being paid for taking on the additional risk.

State intervention and the role of international development agencies is also crucial, especially for emerging countries. While the insurance industry has shouldered losses amounting to USD $1.35 trillion since 1980, that is only one-third of global extreme weather losses—the remainder of which is subsidized by governments or international donors. Governments play a crucial role in improving disaster preparedness at the same time as enabling better disaster prevention and climate adaptation. International organizations such as the World Bank provide assistance and disaster finance for reconstruction and improving the resilience of less developed economies. The financial sector strongly cooperates with these organizations in arranging innovative credit and insurance products, that can help communities and countries recover from extreme natural events, and improve their resilience in the face of climate change.

This book is structured as follows. Chapter 2 provides a description of the short- and long-term economic consequences of natural disasters. The focus will be on the main theoretical approaches that adopted to measure how external shocks like natural disasters impact accumulation of physical and intangible capital, productivity, and growth. The empirical evidence on the effects of natural disasters on the economy is abundant and emphasizes the importance of various factors in significantly moderating this relationship. Moreover, the findings are not uniform across different geographic regions and countries with different levels of pre-disaster economic development.

Chapter 3 looks at how banks are affected by natural disasters in terms of risk, profitability, and the evolving regulatory framework related to climate risks for banks. The channels through which banks are affected by disasters will be highlighted, and a detailed view of the current empirical literature will be provided. Its findings suggest that many external and internal factors to these institutions influence the link between weather hazards and bank activity. A strong emphasis will be dedicated to the regulatory framework shaping bank activity in the face of the climate transition context in different countries.

In Chapter 4, the role of banks in promoting growth through the lending channel from a theoretical and empirical perspective is outlined. Numerous economic theories have attempted to explain the bank lending channel and banks' impact on the real economy. What is different now is that banks' response to natural disasters has a tangible impact on the recovery of the regions where they operate, conditional on other factors, which could be bank, or country specific. Finally, a possible interpretation of how bank strategies and lending could be shaped in the following years by these rules will be provided, conditional on the different climate environment where they will operate and on the regulatory framework they must comply with.

Last, in Chapter 5 the role of other institutions in increasing resilience from natural disasters is described. These could be multilateral development institutions such as the World Bank, the International Monetary Fund, or insurance companies that play a key role in absorbing losses deriving from weather hazards. The role of these organizations in assisting less developed countries in their adaptation efforts to climate change cannot be understated. At the same time, the role of the government is fundamental for building resilience to natural disasters and promoting a culture of risk management in our societies.

REFERENCES

Blickle, K.S., Hamerling, S.N., Morgan, D.P., 2022. How bad are weather disasters for banks? Federal Reserve Bank of New York Staff Reports, No. 990.

Bolton, P., Kacperczyk, M., Hong, H., Vives, X., 2021. Resilience of the Financial System to Natural Disasters. Centre for Economic and Policy Research, London, UK.

Hosono, K., Miyakawa, D., Uchino, T., Hazama, M., Ono, A., Uchida, H., Uesugi, I., 2016. Natural disasters, damage to banks, and firm investment. International Economic Review, 57, pp. 1335–1370.

Odell, K.A., Weidenmaier, M.D., 2004. Real shock, monetary aftershock: The 1906 San Francisco earthquake and the panic of 1907. Journal of Economic History, 64, 1002–1027.

CHAPTER 2

The Economic Impact of Natural Disasters

Abstract Statistics indicate that recorded disaster events and losses have steadily increased in the last 30 years. Researchers have dedicated considerable attention to the economic impact of natural disasters, especially their long-term consequences. This literature acknowledges that the impact of natural disasters on growth is always mediated by other socio-economic factors related to a single country or region, such as the quality of the institutions, political stability, financial sector development, pre-disaster income per capita, or financial liberalization.

Keywords Economic growth · Vulnerability and resilience to disasters · Emerging and developed countries · Socio-economic determinants

2.1 Economic Costs of Natural Disasters

There are many formal definitions of natural disasters. One of them, offered by Encyclopedia Britannica, is the following: Natural disasters can be any calamitous occurrence generated by the effects of natural, rather than human-driven, phenomena that produce significant loss of human life or destruction of the natural environment, private property, or public

A. Duqi, *Banking Institutions and Natural Disasters*, Palgrave Studies in Impact Finance, https://doi.org/10.1007/978-3-031-36371-9_2

infrastructure. Therefore from an economic perspective, natural disasters are events that negatively impact on output, employment, or assets.

Statistics indicate that recorded disaster events have increased in the last 30 years. Disaster losses have been increasing as well. However, there is a considerable year-to-year variation, and across different types of countries (high-middle income vs. less developed ones).

Figure 2.1 shows the damage caused by different types of natural disasters since 1949 collected by EM-DAT[1] EM-DAT divides natural catastrophic events into different groups, meteorological (extreme temperatures, storms, hurricanes), hydrological (landslide, floods, etc.), climatological (wildfires, droughts, etc.), and geological (volcanoes, earthquakes, etc.).[2]

Table 2.1 reports total damages in billions of USD in each macro geographic area for the 1949–1980 and 1981–2022 periods, respectively. Once again, the numbers indicate that each area has experienced a massive increase in reported damages from natural disasters except in a few cases.

The costliest disasters in 2021 were storms, followed by floods, droughts, and earthquakes.[3] If the focus is on disasters' distribution across different types of countries (high vs. low income), Fig. 2.2 shows that most of the losses in absolute terms are endured by high-income countries.

Regarding the number of killed or affected people, it can be observed that there is a large variability in the data. The number of people killed by natural disasters has been increasing even in developed regions such as Western Europe (Tables 2.2 and 2.3). Conversely, the number of affected people per 100,000 inhabitants is much lower in more developed economies, such as Western Europe and North America, compared

---

[1] The EM-DAT database is maintained by the Centre for Research on the Epidemiology of Disasters (CRED), at the Université catholique de Louvain, Belgium. EM-DAT distinguishes between three generic disaster categories: natural, technological, and complex. The natural disaster category is divided into five sub-groups (geophysical, meteorological, hydrological, climatological, biological, and extra-terrestrial), covering 15 disaster types and more than 30 sub-types. For a disaster to be entered into the database, at least one of the following criteria must be fulfilled: (a) Ten (10) or more people reported killed, (b) Hundred (100) or more people reported affected, (c) a Declaration of a state of emergency, (d) Call for international assistance.

[2] In this analysis, biological, technological, and complex disasters are excluded.

[3] EM-DAT CRED, 2021 and Statista 2023 database.

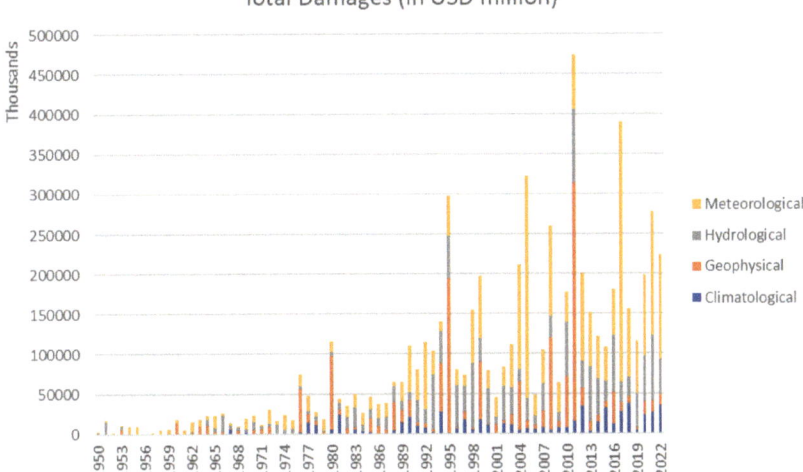

Fig. 2.1 Total damages in USD million generated by different types of natural disasters in the period 1949–2022 (*Source* Author's calculation based on EM-DAT CRED/UC Louvain, Brussels, Belgium, www.emdat.be)

to Asia Pacific, Latin America, or Sub-Saharan Africa. There is a large dispersion across each country grouping, though.

These outcomes are confirmed by the data presented in Table 2.4, which shows the number of deaths per 100,000 population of countries grouped by their Social Development Index (SDI). Less developed countries are more vulnerable to severe natural disasters. Large spikes in death rates occur almost exclusively for countries with a low or low-middle SDI. Highly developed countries are much more resilient to disaster events and have a consistently low death rate from natural disasters.

Table 2.4 indicates that fatalities from natural disasters are exceptionally high when catastrophic events strike countries belonging to the lower-median SDI cohort. Therefore, they are most vulnerable to particularly severe rare weather events.

The measurement of losses after a natural disaster has been argument of debate for academia. First, this could derive from the adoption of

Table 2.1 Total damages in USD billion for different world regions

Region	Climatological		Geophysical		Hydrological		Meteorological	
	1949–1980	1981–2022	1949–1980	1981–2022	1949–1980	1981–2022	1949–1980	1981–2022
Asia and the Pacific	3.30	121.20	47.75	840.70	33.80	872.80	37.32	554.50
Australia	5.88	37.32	0.04	39.86	0.75	35.77	12.73	32.93
Eastern Europe	–	12.81	12.80	41.15	4.68	48.35	–	6.88
Latin America and the Caribbean	10.97	41.20	35.40	108.80	7.93	86.07	24.85	239.20
MENA	0.10	3.78	20.75	61.47	2.77	19.29	0.02	11.36
Northern America	15.24	224.80	17.18	79.75	30.44	154.80	134.80	1,567.00
Sub-Saharan Africa	3.76	10.76	0.21	1.20	3.13	17.14	4.24	10.65
Western Europe	5.33	59.33	109.90	75.63	37.99	233.20	20.93	212.30

Autor's calculation based on EM-DAT CRED/UC Louvain, Brussels, Belgium, www.emdat.be

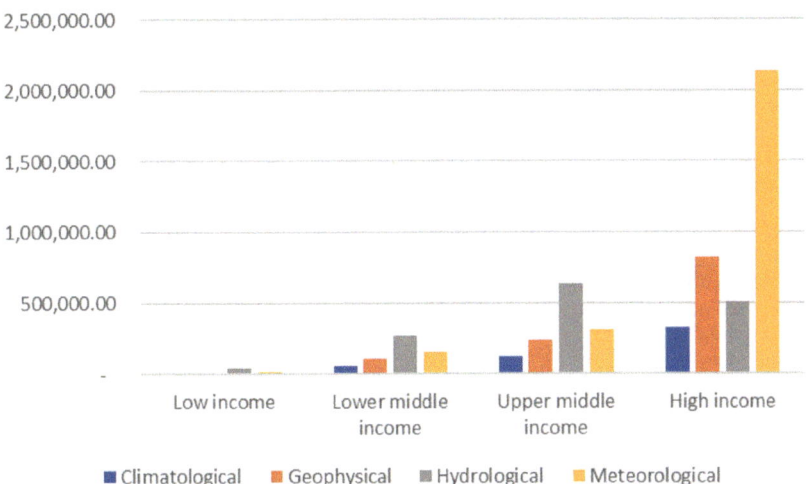

Fig. 2.2 Damage losses in the 1981–2022 period (in USD million) (*Source* Author's calculation based on EM-DAT CRED/UC Louvain, Brussels, Belgium, www.emdat.be)

Table 2.2 Median fatalities per 100,000 inhabitants per region (1981–2022)

Region	Median fatalities per 100,000 inhabitants	25% percentile	75% percentile	Standard deviation
Sub-Saharan Africa	0.574	0.209	1.773	386.57
Asia and the Pacific	0.139	0.032	0.675	66.01
Australia and New Zealand	1.171	0.056	0.460	4.192
Eastern Europe	0.282	0.095	0.912	22.19
Latin America and the Caribbean	0.519	0.156	2.158	621.31
Middle East and North Africa	0.495	0.187	1.245	388.98
Northern America	0.030	0.010	0.077	2.64
Western Europe	0.190	0.066	0.738	32.66

Author's calculation based on EM-DAT CRED/UC Louvain, Brussels, Belgium, www.emdat.be

Table 2.3 Median number of affected people per 100,000 inhabitants per region (1981–2022)

Region	Median number of affected per 100,000 inhabitants	25% percentile	75% percentile	Standard deviation
Sub-Saharan Africa	104.707	12.916	987.024	10,335.54
Asia and the Pacific	16.357	1.561	211.421	6,299.48
Australia and New Zealand	7.597	2.102	39.068	3459.21
Eastern Europe	4.974	1.041	34.065	5080.01
Latin America and the Caribbean	50.920	5.973	417.228	6886.31
Middle East and North Africa	16.747	1.050	253.592	5767.09
Northern America	0.326	0.058	2.436	1143.94
Western Europe	4.117	0.541	32.135	5092.63

Author's calculation based on EM-DAT CRED/UC Louvain, Brussels, Belgium, www.emdat.be

Table 2.4 Average number of deaths per 100,000 inhabitants for different levels of Social Development Index (1981–2022)

Region	Mean all disasters	Mean large disasters
Low SDI	0.018	0.017
Lower-Median SDI	0.014	0.056
Upper-Median SDI	0.0057	0.007
Upper SDI	0.0065	0.005

Author's calculation based on EM-DAT CRED/UC Louvain, Brussels, Belgium, www.emdat.be and IISS-Erasmus University of Rotterdam

different methodologies and metrics, that generate frequent discrepancies among the damage costs reported by different sources. Downton and Pielke (2005) show that loss estimates reported by different sources differ by a factor of 2 or more for the majority of floods that cause less than $50 million in damages.

Second, disaster costs are commonly classified into two broad groups, direct and indirect, and the latter are more complex to evaluate (Pelling et al. 2002; Hallegatte 2015). Houses destroyed by the wind, roads or bridges washed away by floods, injuries, and deaths in the population are typical examples of direct costs.[4] Indirect losses span over a long term. They are consequences of the event, though not directly provoked by it. Examples of these could be a fire caused by an earthquake or loss of tax revenues for the government due to reduced economic activity.

Other authors adopt a different cost terminology. They divide disaster consequences into asset losses (reduction in stock values) and output losses (reduction in income flows) (Rose et al. 2007; Hallegatte 2015). The latter comprise business interruptions, macro-economic effects (reduced demand from customers and businesses), production losses, supply chain disruptions, and long-term effects on growth (part of the population decides to leave the affected area because of a higher risk perception), or higher production due to reconstruction activities.

2.2 WHAT DOES ECONOMIC THEORY SAY ABOUT THE POST-DISASTER GROWTH PATH?

The economic impact of natural disasters has attracted considerable attention from researchers. Most of the research, though, has been devoted to estimating long-term indirect losses from natural disasters, given the difficulty in identifying and measuring them and the underlying mechanisms or moderating factors. The models that aim to estimate this nexus convert the economic reality into a mathematical representation of the most important causal relationships between different economic variables after the shock generated by a natural disaster. In the literature, there are different families of these models, which we briefly summarize here.[5]

2.2.1 *Neo-Classical Growth Models*

These models are based on the neo-classical growth model of Ramsey-Solow-Swan (RSS). It can be summarized as follows:

$$Y = K^{\alpha}(AL)^{1-\alpha}$$

$$K_{t+1} = K_t - \delta K_t + I_t$$

---

[4] EM-DAT reports only direct costs related to natural disasters.

[5] There are several reviews of this literature. See Cavallo and Noy (2009) and Wouter Botzen et al. (2019).

$$I_t = sY_t$$

where Y represents output, A is an index of labor augmenting productivity, K is the capital stock, L is labor, and α is the output elasticity of capital (Wouter Botzen et al. 2019). The rate at which capital depreciates over time is δ, and I is the investment through which capital accumulates. In a steady-state equilibrium, I is equal to savings (s), which are a percentage of Y.

The economy is assumed to be in a steady-state equilibrium. A natural disaster will modify the capital stock or the labor supply, and, therefore shifting it out of this state. A gradual return to the steady state is expected unless disasters permanently affect productivity growth, depreciation, or the savings rate.

2.2.2   Models with Endogenous Productivity

The RSS models described above assume an exogenous and given productivity. Endogenous models aim to address this limitation. Assuming that the goal is to assess the impact of disasters on growth per capita, the model could be estimated through a Cobb–Douglas production function:

$$y_t = A_t^{1-\alpha-\beta} * h_t^{\beta} * k_t^{\alpha} \ \ 0 < \alpha, \ \beta < 1$$

where A_t is the level of available technology, h_t and k_t are respectively human and physical capital–labor ratio, y_t is output produced per capita at time t (Sala-i-Martin and Barro 1995; Klomp and Valckx 2014). Natural disasters affect technological progress, but their impact is debatable. The impact on technology growth is negative when disasters destroy valuable know-how, training facilities, and R&D. Rather, the impact on productivity could be positive when an obsolete technology is replaced by a more advanced one (Hallegatte and Dumas 2009). Chhibber and Laajaj (2008) outline four alternative scenarios of natural disasters' impact on economic growth (Fig. 2.3).

All models predict a drop in the GDP in the aftermath of a disaster due to the destruction of production capacities. The negative impact on GDP is followed by an expansion in scenarios A and B. An inflow of FDI, or foreign aid, could stimulate a higher return on capital and savings. The capital–labor ratio increases due to reconstruction investments. Overinvesting pushes the GDP growth above its natural long-term

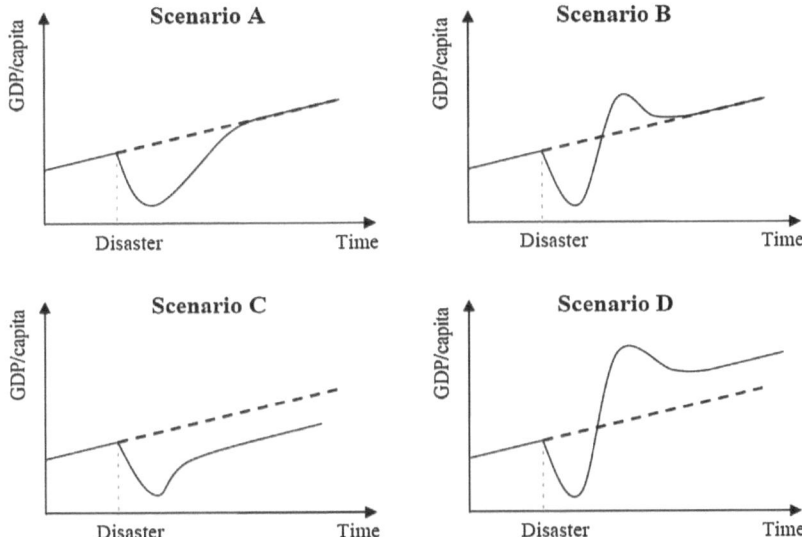

Fig. 2.3 Different scenarios for the long-run consequences of natural disasters on growth (*Source* Chhibber and Laajaj 2008)

track in scenario B. However, income remains in its long-term path due to a larger depreciation of capital compared to replacement investment (Albala-Bertrand 1993; Klomp and Valckx 2014).

If countries cannot return to their previous long-term growth track because financial constraints force economic actors to underinvest, the capital–labor ratio will inevitably decrease (Benson and Clay 2003) (Scenario C). Finally, scenario D hypothesizes that severe disasters can accelerate capital upgrading with more productive versions, and this technological change increases productivity (Skidmore and Toya 2002). In more recent models, productivity growth is driven by economic agents who decide to allocate scarce resources to knowledge creation (Cuaresma 2010; Wouter Botzen et al. 2019).

2.2.3   Input–Output Models and Computable General Equilibrium Models

Through these models, researchers look at affected companies within a regional or national economy after a natural disaster. They aim to capture interdependencies between sectors that are part of the firm's supply chain and subsequently try to estimate the so-called "Ripple effect"; how the loss in a particular area spills over to other regions or sectors (McCarty and Smith 2005). The main concern with these models is the assumption of an invariant technology and behavior of economic agents. Therefore they fail to account for dynamic adjustment processes after a disaster strike (Okuyama 2004; Hallegatte 2008).

2.3   EMPIRICAL EVIDENCE FROM EMERGING AND DEVELOPED ECONOMIES

Considerable research has been devoted to the consequences of natural disasters on economic growth, with a particular focus on developing economies, since they seem to suffer more from these shocks. Looking first at cross-country studies, what emerges is that the findings are not conclusive. Albala-Bertrand (1993) is one of the first contributions to this literature. He investigates the long-run impact of several catastrophic weather events during the period 1960–1979 in a sample of Latin and Central American countries. Localized natural disasters do not significantly affect these countries' long-term growth or inflation rates, and much of the impact that could be erroneously attributed to the disaster is related to other domestic ongoing factors. The disaster situation in each country comingles with social and political developments.

Skidmore and Toya (2002) build an extensive dataset spanning from 1960 to 1990 for 89 countries and estimate the impact of disasters on a series of economic indicators such as human capital accumulation, physical capital investment, total factor productivity, and growth. The results show that climatic (geologic) disasters are positively (negatively) associated with growth. Moreover, disasters impact growth by increasing human capital investment rates and total factor productivity, but they do not affect physical capital accumulation. Disasters provide opportunities for an update of the capital stock and adoption of new technologies.

Cuaresma et al. (2008) focus on the impact of disasters on foreign technology absorption in developing economies. The latter fail to benefit from technology spillovers from more advanced countries. These findings are, therefore not consistent with the "creative destruction" role that Skidmore and Toya (2002) seemed to attribute to natural disasters. When countries in the sample are divided by the overall level of development, the results indicate that technology spillovers after natural catastrophes are positive only for those in the highest deciles of the GDP distribution per capita.

Noy (2009) examines the impact of disaster damage on long-term growth for developing economies. He obtains robust evidence that the amount of property damage incurred during the disaster is a negative determinant of post-disaster GDP growth performance. The reason for this is that the adverse impact of a disaster is caused mainly through damage to the capital stock, delivery and transportation systems, and other infrastructure. There is also evidence that the most significant indirect costs from disasters are borne by developing countries, even when disasters are of similar magnitude to those experienced by more advanced economies.

Loyaza et al. (2012) explore the effects of natural disasters on growth separately by disaster and economic sector. They apply a dynamic generalized method of moments panel estimator to a 1961–2005 cross-country panel dataset. Three significant insights emerge. First, economic growth is not always negatively affected by natural disasters. Their consequences differ by types of disasters and economic sectors. Second, although moderate disasters (such as moderate floods) can have a positive growth effect in some sectors, severe disasters do not. Third, growth in developing countries is more sensitive to natural disasters, with more sectors affected, and larger and more economically meaningful effects. Fomby et al. (2013) reach to similar results after disaggregating gross domestic product into agricultural and nonagricultural components, differentiating between advanced and developing countries, and severe vs. moderate disasters.

Hochrainer (2009) and Cavallo et al. (2013) further elaborate on the impact of natural disasters on growth in a cross-country sample. They reach to a similar conclusion as Albala-Bertrand (1993) to the extent that large catastrophic disasters affect short and long-run growth, but this could be due to the confounding effect of existing political instability and radical political revolutions following disasters.

Cunado and Ferreira (2014) focus solely on floods on a novel dataset spanning 135 countries. By adopting a panel VAR technique, they evidence that flood shocks tend to have positive impacts on GDP growth rates. The growth is not materialized in the year of the flood but later, considering that the growth response of floods in the agricultural sector has beneficial effects on land productivity that manifest during subsequent harvest cycles. Developing countries, which typically rely on more traditional, less intensive forms of agriculture, benefit from a more consistent increase in agricultural growth in the years following the floods.

In a similar setting, Berlemann and Wenzel (2016) look at the effect of droughts on a large sample of 153 countries over the period 1960–2002. They document significantly adverse long-term growth effects of droughts in both highly and less developed countries. Droughts depress human capital formation through a decrease of the years of total schooling, which leads to a drop of the savings rate of the population in the medium term.

More recently, Budina et al. (2023) find that natural disasters exacerbate inequality and poverty rates in advanced and emerging economies. More surprisingly, inequality increases more in developed economies after disaster occurrences. This effect could be attributable to a lack of insurance coverage in emerging markets, which provokes worsened economic conditions for everyone after the disaster strike.

As natural disasters are likely to become more frequent because of climate change, the empirical outcomes suggest that inequality will exacerbate in all countries regardless of their level of development.

Several studies have focused on single countries/regions and the impact that particular types of natural disasters generate at the local level or regional level. Belasen and Polachek (2009) look at the local impact of hurricanes on workers' wages in Florida using a DID econometric approach. They are among the first to consider the regional/local effects of extreme weather events compared to previous studies that adopted a macro view. The results highlight a decreased employment and higher wages in counties stricken by hurricanes. These hazards appear to negatively impact labor supply while at the same time changing the labor demand for specific industrial sectors. As workers flee the devastation by heading into neighboring counties, those counties experience a positive labor supply shock. The result is that employment will remain relatively unchanged while earnings decline.

Strobl (2011) is notably one of the few studies that looks at the local economic impact of hurricanes in the U.S. by looking at county level data for the period 1970–2005. A hurricane landfall in a county reduces its per capita income growth rate by 0.45 percentage points, compared to the average growth rate of U.S. counties of 1.68%. Reduction in per capita income originates from an endogenous mobility response to the hurricane, whereby more affluent individuals are more likely to migrate out of the affected county.

The consequences of a large disaster on local labor markets are also the focus of Kirschberger (2017). She explores how a large earthquake, which hit the capital of Indonesia in 2006, affected labor market outcomes, particularly the evolution of wages across sectors. Her findings are aligned with the "creative destruction" theory illustrated above. Labor market outcomes for individuals living in earthquake-affected regions proved remarkably resilient to such large shocks, and on average, wage growth did not vary significantly. Individuals working in agriculture sectors experienced higher growth in earnings, because these markets were segmented and mobility was rather limited.

Noy and Vu (2010) study the short-run effects of natural disasters in Vietnam, a developing country. The results indicate that extreme weather events have a positive impact on growth. For a one percentage point increase in the ratio of direct damage to output, output growth increases by 0.04%. These results are driven by those regions of the country that have more access to reconstruction fund, and high levels of development, which allows them to enjoy capital upgrading after the occurrence of natural disasters.

Hsiang (2010) studies the effect of cyclone intensity on economic activity in 28 Caribbean countries. Cyclones' impact on overall GDP is relatively small, but when this effect is decomposed by the industrial sector, both significant negative and sizeable positive output responses emerge. The only sector that benefits from this type of natural disaster is construction, presumably because of reconstruction efforts, whereas tourism, agriculture, and trade are all impacted negatively.

Anttila-Hughes and Hsiang (2013) examine tropical cyclones and how they affect household income in the Philippines. They find that typhoons affect all households (rich and poor alike) and that annual income per household is reduced by 6.6% in the aftermath of a disaster. This leads to a reduction in household expenditures related to human capital investment.

Mohan et al. (2018) investigate the impact of hurricanes on 21 Caribbean countries for the period 1970–2011. They disentangle how each national income component, i.e., export, import, public consumption, private investment, and private consumption responds to natural disaster shocks. The main findings indicate a slight increase in the GDP after the hurricane followed by a later contraction, which stems mainly from a temporary expansion of import, investment, and government expenditures, which tend to reverse in the medium term.

Ishizawa and Miranda (2019) document a short-term reduction of GDP per capita, income, and an increase of poverty rates in Central America following hurricane strikes in these countries. Friedt and Toner-Rogers (2022) document significant and persistent investment reductions across affected Indian regions following a disaster, and lasting positive spillovers into otherwise unaffected areas. Private investment from multinational firms tends to flow into more developed, less disaster-prone regions, fueling the prominent divergence in India's economic growth and suggesting that large corporations are including climate risks into their strategies.

Joseph (2022) examines the average causal impact of the 2010 earthquake in Haiti on economic growth and recovery by conducting a disaggregated empirical analysis at the sub-national level. The results indicate that the earthquake negatively affected the country's economic growth and that this decline in growth persisted up to ten years after the disaster.

Leiter et al. (2009) offer an innovative view by examining the impact of floods on a series of corporate economic indicators for a sample of European firms. Differently from other studies, they focus only on the short-run effects of flood effects on European businesses and differentiate between firms with higher or lower tangible and intangible capital. The econometric implementation follows a *difference-in-differences* (DID) approach. The results show that firms with a high pre-disaster percentage of intangibles in the asset mix increase productivity at a higher rate than others, which rely more on tangible assets.

2.4 COUNTRY DETERMINANTS OF VULNERABILITY AND RESILIENCE TO NATURAL HAZARDS

The empirical evidence presented in the previous section has generally acknowledged that the impact of natural disasters on growth is always mediated by other socio-economic factors related to a single country or region. The reason is that vulnerability to disasters, the human footprint on the environment, and development are closely interlinked. For instance, many have pointed out that the massive damage caused by floods in Pakistan in 2022 was attributable to climate change, and to the largely inadequate infrastructure to absorb climate shocks.

In general, how countries recover after a severe weather event can be related to their level of development, financial systems, the quality of the institutions, size, and other external factors. A large empirical evidence has shown that disasters of similar magnitude impact differently advanced rather than developing countries, and the damage toll is considerably higher for the latter. This difference is likely due to the greater amount of resources spent on prevention efforts and legal enforcement of mitigation rules (e.g., building codes) in more developed economies. In particular, some of the policy interventions likely to ameliorate disaster impact, including land-use planning, building codes, and engineering interventions, are rare in less developed countries (Cavallo and Noy 2009). Nations can successfully address weak institutions, create better insurance markets, require more stringent building standards, reduce corruption, and institute more advanced warning and emergency response systems when levels of development have reached a certain point (Kellenberg and Mobarak 2008).

Country size and geographic location are determinants of vulnerability to natural disasters. Bigger economies are more diversified and capable of engineering the inter-sectoral and inter-regional transfers required to mitigate the economic impact of natural disasters. The geographic position of certain countries or regions increases their exposure to specific types of natural disasters. Coastal areas located nearby tropical seas are greatly affected by tropical cyclones. The small island states of the Caribbean region are particularly vulnerable on both dimensions (Rasmussen 2004; Heger et al. 2008). In contrast, even by their size alone, large developed countries can more easily absorb output shocks from natural disasters hitting a few regions of the country (Auffret 2003).

Extensive research has focused on the political and institutional factors that affect disasters' impact. A consistent finding of several studies (i.e., Kahn 2005; Skidmore and Toya 2007; Raschky 2008) is that better institutions, greater accountability of the government, more stable democratic regimes, or greater security of property rights reduce disaster impact. Anbarci et al. (2005) elaborate on the political economy of disaster prevention. They conclude that unequal societies cannot implement preventive and mitigating measures, as they cannot take decisions that benefit the collectivity. Therefore, unequal societies tend to have fewer resources spent on prevention.

Similarly, Besley and Burgess (2002) observe that flood impacts in India are negatively correlated with newspaper distribution; when circulation is higher, politicians are more accountable and the government is more active in preventing and mitigating the impacts of disasters. Eisensee and Strömberg (2007) reach similar conclusions regarding the U.S. disaster aid response to media reports.

Felbermayr and Gröschl (2014) find that democratic countries can better cope with natural disasters than autocracies. Moreover, countries that are more open to international trade absorb the shock of a disaster better because replenishing of the capital stock is sped up by the availability of foreign funds and investment goods.

Noy (2009) divides these factors into domestic policy, structural and external ones. The first group comprises changes in government budget surpluses, inflation, investment, and credit growth. External factors include the current account and foreign direct investment inflows, the structural factors are the openness of the economy to international trade (import ratio), a measure of institutional quality, and a binary measure for financial crises. He finds that developing and small countries face much larger shocks to their macroeconomies compared to developed and large ones following a disaster of similar relative magnitude. Countries with higher literacy rates, better institutions, higher per capita incomes, larger governments, and a higher degree of openness to trade are more able to withstand the initial disaster shock and prevent its effects spilling deeper into the macroeconomy. Financial conditions also seem to matter. Countries with fewer open capital accounts, more foreign exchange reserves, and higher levels of domestic credit appear more robust and able to endure natural disasters with less spillover to GDP growth rates.

Cuaresma et al. (2008) show that countries with higher GDP per capita can absorb technology spillovers. Loyaza et al. (2012) find that the proxies of initial educational investment, depth of financial intermediation, and trade openness reduce the negative impact of disasters on economic growth. At the same time, macroeconomic price instability, significant fiscal burdens, and unfavorable terms of trade negatively impact on post-disaster growth rates. They elaborate further on the effect of different types of disasters on different industrial sectors. Floods are found to positively affect the growth of the agricultural sector, differently from storms. Therefore the countries' reliance on specific industrial sectors may affect their post-disaster growth rate. Fomby et al. (2013) follow a similar econometric methodology and reach similar results. The growth response of different sectors of the economy is heterogeneous to different types of natural disasters.

McDermott et al. (2014) link the medium-term growth dynamics in the aftermath of a disaster with the level of financial sector development. In the absence of well-developed financial markets, liquidity constraints may cause income shocks that produce more significant long-term effects. The shock generates destructive effects on lifetime wealth, and this outcome is amplified for poorer households whose future earning potential is also reduced through the forced disinvestment of productive assets. They are forced to divest productive assets in countries with weak financial sector development, further constraining growth prospects.

Another group of papers has looked at aid and disasters from different perspectives, but the research on this topic is still scant and results are ambiguous. Raschky and Schwindt (2016) propose that aid could have unexpected adverse effects on countries' resilience to disasters since it reduces incentives to improve protective measures and increases the likelihood of greater damage in the future. Adam and Bevan (2006) claim that aid can be used to rebuild destroyed infrastructure and resettle displaced workers. Therefore post-shock productivity will recover, and there will be an impetus to economic activity. However, aid could translate into windfall income for some agents, if it targets predominantly the non-tradable sector, boosting wages in that sector and generating a reallocation of (skilled) labor (Corden and Neary 1982). This process more adversely affects the manufacturing sector, since the prices of their output, as traded goods, are fixed.

Bulte et al. (2018) provide an interesting natural experiment of the disaster aid–growth nexus at the regional level following a major earthquake in eastern rural China. They find that industrial GDP fell in the affected provinces after aid inflows, due to the so-called "Dutch disease"; where aid provoked rising prices of non-tradables that undermined the profitability of the manufacturing sector.

Another interesting paper that analyzes the post-disaster growth path at a regional level is Barone and Mocetti (2014). They analyze two separate earthquakes that hit two Italian regions (Friuli and Irpinia) in 1976 and 1980. Their findings show how different regional governments' responses to a major disaster could yield different post-event consequences inside the same country.

Initially, they document no significant effects of the quakes in the short term, which can be primarily attributed to the role of financial aid in the aftermath of the disaster. In the long term, the growth outcomes diverge: the quakes yielded a positive effect in Friuli and a negative one in Irpinia.

In the former, 20 years after the quake, the GDP per capita growth was 23% higher than in the synthetic control, while in the latter, the GDP per capita experienced a 12% drop. The dynamics of the GDP mirror the trend of productivity growth in both regions. The authors provide evidence that institutional quality shapes these results. Since in Irpinia the pre-quake institutional quality was "low" (compared to the national average) while in Friuli it was "high"; the pre-existing local economic and social milieu will likely play a crucial role in the sign of the economic effect of a natural disaster.

The findings of the literature on the disaster–growth nexus and relative moderating factors are summarized in Fig. 2.4. Variables of interest have been grouped into natural disaster types, drivers of GDP growth rates, moderating factors, and GDP components. Moderating factors can be classified into socio-political, economic/financial, or others.

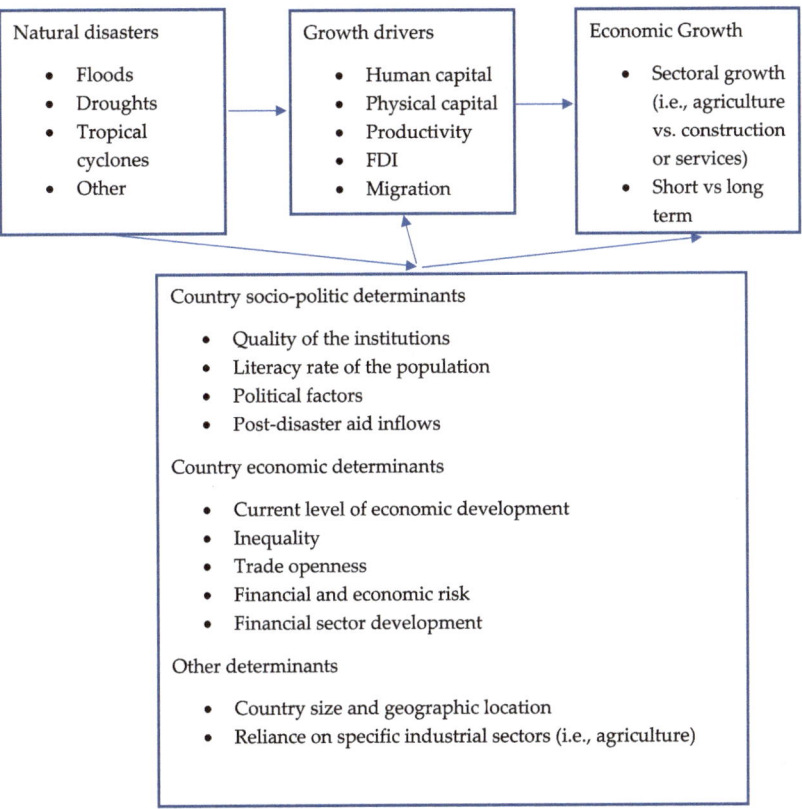

Fig. 2.4 Summary of main findings from the literature on the impact of natural disasters on growth

REFERENCES

Adam, C., Bevan, A., 2006. Aid and the supply side: public investment, export performance, and Dutch disease in low-income countries. World Bank Economic Review, 20, pp. 261–290.

Albala-Bertrand, J., 1993. Natural disaster situations and growth: A macroeconomic model for sudden disaster impacts. World Development, 21, pp. 1417–1434.

Anbarci, N., Escaleras, M., Register, C.A., 2005. Earthquake fatalities: The inter-action of nature and political economy. Journal of Public Economics, 89, pp. 1907–1933.

Anttila-Hughes, J.K., Hsiang, S.M., 2013. Destruction, disinvestment, and death: Economic and human losses following environmental disaster. Available at SSRN: https://ssrn.com/abstract=2220501 or http://dx.doi.org/10.2139/ssrn.2220501.

Auffret, P., 2003. High consumption volatility: The impact of natural disasters? World Bank Policy Research Working Paper, No. 2962.

Barone, G., Mocetti, S., 2014. Natural disasters, growth and institutions: A tale of two earthquakes. Journal of Urban Economics, 84, pp. 52–66.

Belasen, A., Polachek, S.W., 2009. How disasters affect local labor markets? The effects of hurricanes in Florida. Journal of Human Resources, 44, pp. 251–277.

Benson, C., Clay, E., 2003. Disasters, vulnerability and the global economy. In: A. Kreimer, M. Arnold, and A. Carlin (eds.), Building Safer Cities: The Future of Disaster Risk, pp. 3–31.

Berlemann, M., Wenzel, D., 2016. Long-term growth effects of natural disasters—Empirical evidence for droughts. Economics Bulletin, 36, pp. 464–476.

Besley, T., Burgess, R., 2002. The political economy of government responsiveness: Theory and evidence from India. Quarterly Journal of Economics, 117, pp. 1415–1451.

Budina, N., Chen, L., Nowzohour, L., 2023. Why some don't belong? The distributional effects of natural disasters. IMF Working Paper Series, No. 23/2.

Bulte, E., Xu, L., Zhang, X., 2018. Post-disaster aid and development of the manufacturing sector: Lessons from a natural experiment in China. European Economic Review, 101, pp. 441–458.

Cavallo, E., Galiani, S., Noy, I., Pantano, J., 2013. Catastrophic natural disasters and economic growth. Review of Economics and Statistics, 95, pp. 1549–1561.

Cavallo, E., Noy, I., 2009. The economics of natural disasters: A survey. IDB Working Paper Series, No. 124.

Chhibber, A., Laajaj, R., 2008. Natural disasters and economic development impact: Response and preparedness. Journal of African Economies, 17, pp. 7–49.

Corden, W., Neary, P., 1982. Booming sector and de-industrialisation in a small open economy. Economic Journal, 92, pp. 825–848.

Cuaresma, J.C., 2010. Natural disasters and human capital accumulation. World Bank Economic Review, 24, pp. 280–302.

Cuaresma, J.C., Hlouskova, J., Obersteiner, M., 2008. Natural disasters as creative destruction? Evidence from developing economies. Economic Inquiry, 46, pp. 214–227.

Cunado, J., Ferreira, S., 2014. The macroeconomic impacts of natural disasters: The case of floods. Land Economics, 90, pp. 149–168.

Downton, M.W., Pielke, R.A., 2005. How accurate are disaster loss data? The case of U.S. flood damage. Natural Hazards, 35, pp. 211–228.

Eisensee, T., Strömberg, D., 2007. News floods, news droughts, and U.S. disaster relief. Quarterly Journal of Economics, 122, pp. 693–728.

Felbermayr, G., Gröschl, J., 2014. Naturally negative: The growth effects of natural disasters. Journal of Development Economics, 111, pp. 92–106.

Fomby, T., Ikeda, Y., Loyaza, N.W., 2013. The growth aftermath of natural disasters. Journal of Applied Econometrics, 28, pp. 412–434.

Friedt, F.L., Toner Rodgers, A.T., 2022. Natural disasters, intra-national FDI spillovers, and economic divergence: Evidence from India. Journal of Development Economics, 157, p. 102872.

Hallegatte, S., 2008. An adaptive regional input-output model and its application to the assessment of the economic cost of Katrina. Risk Analysis, 28, pp. 779–799.

Hallegatte, S., 2015. The indirect cost of natural disasters and an economic definition of macroeconomic resilience. Global Facility for Disaster Reduction and Recovery, The World Bank Group.

Hallegatte, S., Dumas, P., 2009. Can natural disasters have positive consequences? Investigating the role of embodied technical change. Ecological Economics, 68, pp. 777–786.

Heger, M., Julca, A., Paddison, O., 2008. Analysing the impact of natural hazards in small economies: The Caribbean case. UNU/WIDER Research Paper, No. 25.

Hochrainer, S., 2009. Assessing the macroeconomic impacts of natural disasters: Are there any? World Bank Policy Research Working Paper No. 4968.

Hsiang, S. M., 2010. Temperatures and cyclones strongly associated with economic production in the Caribbean and Central America. Proceedings in Natural Academic Sciences, 107, pp. 15367–15372.

Ishizawa, O.A., Miranda, J.J., 2019. Weathering storms: Understanding the impact of natural disasters in Central America. Environmental Resource Economics, 73, pp. 181–211.

Joseph, I., 2022. The effect of natural disaster on economic growth: Evidence from a major earthquake in Haiti. World Development, 106053.

Kahn, M.E., 2005. The death toll from natural disasters: The role of income, geography, and institutions. Review of Economics and Statistics, 87, pp. 271–284.

Kellenberg, D.K., Mobarak, A.M., 2008. Does rising income increase or decrease damage risk from natural disasters? Journal of Urban Economics, 63, pp. 788–802.

Kirschberger, M., 2017. Natural disasters and labor markets. Journal of Development Economics, 125, pp. 40–58.

Klomp, J., Valckx, K., 2014. Natural disasters and economic growth: A meta-analysis. Global Environment Change, 26, pp. 183–195.

Leiter, A.M., Oberhofer, H., Raschky, P.A., 2009. Creative disasters? Flooding effects on capital, labour and productivity within European firms. Environmental Resource Economics, 43, pp. 333–350.

Loyaza, N.W., Olaberría, E., Rigolini, J., Christianensen, L., 2012. Natural disasters and growth: Going beyond the averages. World Development, 40, pp. 1317–1336.

McCarty, C., Smith, S., 2005. Florida's 2004 hurricane season: Local effects. Florida Focus, Bureau of Economic and Business Research, University of Florida, 1.

McDermott, T.K.J., Barry, F., Tol, R.S.J., 2014. Disasters and development: natural disasters, credit constraints, and economic growth. Oxford Economic Papers, 66, pp. 750–773.

Mohan, P.S., Ouattara, B., Strobl, E., 2018. Decomposing the macroeconomic effects of natural disasters: A national income accounting perspective. Ecological Economics, 146, pp. 1–9.

Noy, I. 2009. The macroeconomic consequences of disasters. Journal of Development Economics, 88, pp. 221–231.

Noy, I., Vu, T. B., 2010. The economics of natural disasters in a developing country: The case of Vietnam. Journal of Asian Economies, 21, pp. 345–354.

Okuyama, Y., 2004. Modeling spatial economic impacts of an earthquake: Input-output approaches. Disaster Prevention and Management, 13, pp. 297–306.

Pelling, M., Özerdem, A., Barakat, S., 2002. The macro-economic impact of disasters. Progress in Development Studies, 2, 283–305.

Raschky, P.A., 2008. Institutions and the losses from natural disasters. Natural Hazards Earth Systems Science, 8, pp. 627–634.

Raschky, P., Schwindt, M., 2016. Aid, catastrophes and the Samaritan's dilemma. Economica, 83, pp. 624–645.

Rasmussen, T.N., 2004. Macroeconomic implications of natural disasters in the Caribbean. IMF Working Paper, 04/224.

Rose, A., Porter, K., Dash, N., Bouabid, J., Huyck, C., Whitehead, J., Shaw, D., Eguchi, R., Taylor, C., McLane, T., 2007. Benefit-cost analysis of FEMA hazard mitigation grants. Natural Hazards Review, 8, pp. 97–111.

Sala-i-Martin, X., Barro, R., 1995. Economic Growth. MIT Press.

Skidmore, M., Toya, H., 2002. Do natural disasters promote long run growth? Economic Inquiry, 40, pp. 664–687.

Skidmore, M., Toya, H., 2007. Economic development and the impacts of natural disasters. Economic Letters, 94, pp. 20–25.

Strobl, E. 2011. The economic growth impact of hurricanes: Evidence from U.S. coastal counties. Review of Economics and Statistics, 93, pp. 575–589.

Wouter Botzen, W.J., Deschenes, O., Sanders, M., 2019. The economic impacts of natural disasters: A review of models and empirical studies. Review of Environment Economic Policy, 13, pp. 167–188.

CHAPTER 3

The Consequences of Natural Disasters on Banking Institutions

Abstract Extreme weather events threated the banking sector through depreciation of collateral, increase of counterparty bankruptcies, misvaluation of securities held in bank portfolios, and increase of funding risk. At the same time, post-disaster reconstruction boosts the demand for credit, which, if met by banks, can increase their revenues and improve their profitability. Bank regulators around the globe are taking several measures to address climate risk exposure of banks, such as climate stress tests or climate scenario analyses. However, these exercises are still at an early stage and do not adequately quantify the complex relationships between disasters, banks, and the real economy.

Keywords Bank risk and profitability · Disaster risk · Bank stress tests · Banking regulation

3.1 In Terms of Lower Profitability, Resilience, and Risk, Are Banks Affected by Natural Disasters?

In the previous chapter, it was outlined that the impact of natural disasters on the economy is mitigated in the presence of institutions of a higher quality and a well-functioning financial sector. In the current and next chapter, we look more in detail at the role of financial institutions, especially commercial banks, in this context and whether they are affected by climate change and extreme natural hazards.

We start by investigating whether natural disasters impact bank performance. Recently, there has been a renewed interest on the link between catastrophic weather events and the financial sector. A recent IMF report on Caribbean countries claims that more frequent hurricanes could have pronounced effects on the quality of banks' loan portfolios, depending on the country's fiscal, monetary, and prudential policy responses, its dependence on tourism, and the sectoral composition of banks' loan books.[1]

The U.S. Secretary of the Treasury, Janet L. Yellen, recently stated that policymakers need to take action against climate change which threatens the financial sector. In October 2022, the U.S. Treasury established the Climate-Related Financial Risk Advisory Committee of the Financial Stability Oversight Council since climate risks could seriously harm the U.S. financial sector. Similar steps have been taken by other regulators, such as the European Central Bank or the Financial Conduct Authority in the U.K.[2]

However, the potential effects of climate change and acute physical risks such as natural disasters on the banking sector's risk and profitability remain ambiguous and uncertain. Banks have undergone radical changes almost in every country in the world in the last 30 years. After the Global Financial Crisis of 2008–2009 and the Sovereign Debt Crisis of Europe in 2010, the new operating environment has forced banks to reduce risks in order to meet more demanding capital and liquidity requirements and pass stress test requirements set by regulatory agencies (such as the U.S. Federal Reserve and the European Banking Authority).[3] A significant number of systemically important banks which operate worldwide have emerged, and the complexity of activities and interconnections among them has increased dramatically. These giant banks, however, face several

---

[1] Speech by Deputy Managing Director Bo Li's Keynote Speech at the 59th Bi-Annual Meeting of CARICOM Central Bank Governors, Nassau. The text is available here https://www.imf.org/en/News/Articles/2022/11/03/sp110322-financial-sector-risks-and-supervision-in-the-midst-of-climate-change.

[2] Section 3.3 will be devoted to the regulatory framework of climate risks in banking.

[3] Berger et al. (ed.) (2019).

weaknesses, one of which is that the bulk of their deposits is concentrated geographically. This is observed for large U.S. banks for instance, which operate nationwide but fund themselves from deposits originated in few counties. Idiosyncratic shocks such as catastrophic natural disasters in these counties could provoke a permanent decline in deposits. Bank lending would be adversely affected, and growth would be lower. Moreover, under certain conditions, the local shock could propagate to other areas posing a risk to the country's overall economy (Kundu et al. 2022).

There are several channels through which natural disasters and climate risks in general can impact bank stability and profitability. Physical risks arising from catastrophic weather events adversely affect sovereign debt, industrial sectors, corporations, and households. If the focus is at the firm level, acute or chronical physical risks could hamper firm revenues, operational costs, and capex through increased investments to cope with climate change and could reduce asset/equity valuation because of stranded assets. These outcomes could spill over to banks in multiple ways. Depreciation of collateral and counterparty bankruptcies could increase bank risk. A misvaluation of securities that banks hold in their portfolios could increase market risk. Losses due to branch closures and damage caused by disasters could impact on operational risk. If customers withdraw deposits to cope with liquidity needs, the bank could face liquidity issues. Figure 3.1 depicts an overview of these mechanisms.

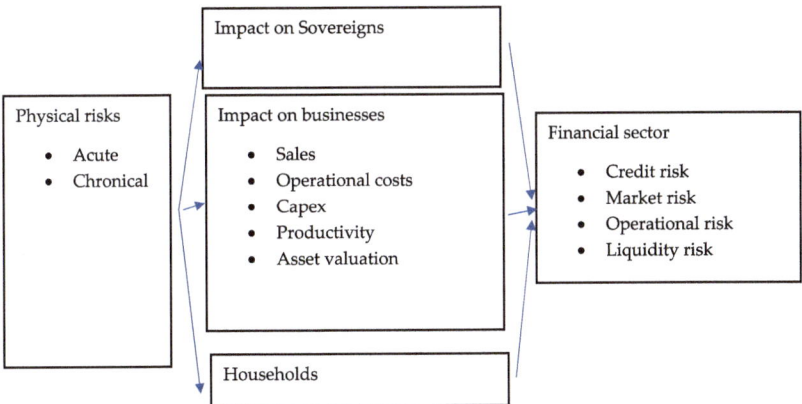

Fig. 3.1 Impact of physical risks on the financial system (Adapted from Network for Greening the Financial System 2020; Bolton et al. 2020a)

In the medium-long term, banks could see the average quality of their customer base deteriorate due to the worsening of economic conditions after the disaster shock. Due to increased climate risks, they could decide to not serve specific segments of the population or the country. This could worsen even more the economic outlook of specific regions.

At the same time, acute events such as hurricanes, floods etc., increase credit demand due to reconstruction needs. This demand, if met by banks, can increase their revenues, boost their margins, and improve bank profitability. Therefore, the impact of natural disasters on the banking sector is a relatively new topic and the consequences on these institutions are still not definitive.

The impact of natural disasters on bank risk and lending will be treated separately in this study. However, the two are strictly interconnected, as we will document below. Regarding bank risk, prior research has tried to differentiate between credit risk, which is captured through an increase of non-performing loans (NPLs), loan loss provisions (LLPs), or regulatory capital ratios. There are also contributions that look at bank systemic risk, liquidity risk, or banks' probability of default if they are exposed to losses from severe weather events.

Albuquerque and Rajhi (2019) look at a large panel of developing countries for the period 1995–2011. They find that the impact on the banking sector can be very detrimental when disasters are associated with country fragility. NPLs temporarily increase after the disaster, as does the likelihood of bank defaults in the medium term, through a depletion of bank reserves and an increase in leverage. Securitization structures could introduce additional risk through interconnections between financial institutions. When catastrophes strike, the transfer of risks to other financial institutions commits banks and their counterparties to pay out when the transferred risk materializes (Von Dahlen and Von Peter 2012).

Collier et al. (2011) look at the effect of El Niño event in 1997–1998 in a poor region of Peru. It generated significant losses to agricultural production and extensive damage to the infrastructure. These disruptions resulted in cash flow problems in local enterprises and consequently affected the rural lenders supporting the latter, especially those with a concentrated loan portfolio in that specific region. NPLs and restructured loans increased almost seven times compared to the pre-disaster period due to the correlated risk exposure of many borrowers.

A recent study by Anderlik et al. (2022) evidences that local banks that concentrated lending in poor counties which were stricken by Hurricane

Katrina were adversely affected by this event. Past-due and nonaccrual loan ratios spike, compared to banks operating in non-poor affected counties. This outcome indicates that the pre-disaster economic development of a specific area plays a determinant role in post-disaster banking performance.

From a disaster risk management perspective, banks can cautiously increase loan loss provisions (LLPs) or tighten credit standards. Dal Maso et al. (2022) report that bank managers in the U.S. observe the components of disaster risk that impact their lending portfolios and estimate adequate LLPs to build reserves to absorb potential future losses. Local banks in the U.S. have increased LLPs proactively by recognizing their exposure to a specific type of disaster. This policy allows them to be prepared for future write-offs by accelerating the recognition of potential future bad debts that otherwise would be recognized in subsequent periods.

Credit standard tightening is also evidenced in Duanmu et al. (2022) and He Huang et al. (2022). The former find that bank residential mortgage lending standards are affected by risks to the local economy from natural disasters. Banks tighten lending standards in disaster-hit counties, suggesting that lenders are more cautious in these locations since environmental disasters can increase the long-term risks to the local economy. The empirical evidence suggests that disaster risk awareness among lenders has increased, and disasters are impacting bank lending standards.

He Huang et al. (2022) examine how a firm's exposure to climate risk affects its financing terms from bank loans. Climate risk exposure is assessed by firm managers and reflects the degree to which the firm is subject to climate-induced natural disasters. The results show that firms face higher interest rates and more stringent collateral and covenant constraints when borrowing from banks if they are exposed to higher climate risk, which hurts firms' financial performance and heightens their default likelihood.

Natural disasters can impact on bank soundness and stability as well. Several papers have looked at the post-disaster bank Distance to Default and bank systemic risk. The Distance to Default is approximated by the number of standard deviations that a bank's return on assets has to drop below its expected value before equity is depleted and the bank is insolvent (Klomp 2014). Adopting this indicator provides a more comprehensive picture of the impact of natural disasters on the banking sector compared to variables that capture only credit risk exposure, such as NPLs. Even in this case, the empirical outcomes reveal differences between developing and advanced economies. More developed financial systems can mitigate the impact of catastrophic events through adequate diversification strategies, a well-functioning insurance market, financial aid provided by government agencies, and appropriate banking regulation, which induces banks to build adequate reserves against future disaster shocks.

Klomp (2014) shows that large-scale natural disasters significantly increase the banking system's fragility. His results support the current opinion that regulators should start implementing bank capital adequacy ratios against potential losses from catastrophic weather events. He also points out that in presence of a well-functioning regulatory sector, the impact of disasters on bank fragility is more attenuated. This outcome is in line with previous research reported in previous sections, which finds that institutional quality is a significant driver of natural disasters' economic impact. Further, disasters impact more on bank stability in emerging economies and in countries with less developed financial markets.

Natural disasters matter for bank stability even in developed economies which are more disaster-prone than others. Noth and Schüwer (2023) find that weather-related natural disasters significantly weaken the stability of banks with business activities in affected regions of the U.S. In particular, banks' probabilities of default increase, z-scores decrease, non-performing asset and foreclosure ratios increase, and returns on assets and equity ratios decrease within two years after the disaster. Bank borrowers are not sufficiently protected against financial difficulties by insurance payments and public aid programs. Therefore their credit quality deteriorates with a negative impact on bank stability. Similar outcomes are evidenced by Apergis (2022).

However, Blickle et al. (2022) reach opposite conclusions. They look at a similar sample as Noth and Schüwer (2023) and Apergis (2022), that is, banks operating in the U.S. counties hit by major disasters over the period 1995–2018. They find that loan losses and default risk increase at local banks (banks operating in a single state), but these adverse outcomes are often offset by local banks' income due to reconstruction loans. So natural disasters do not appear to threaten bank stability.[4]

A number of studies have looked at the possibility that natural disasters might adversely affect bank liquidity risk. Banks could face liquidity pressure if depositors redeem their claims *en masse* due to reconstruction needs, and a disruption in the financial markets, making it difficult for banks to raise funds. This was initially hypothesized by Steindl and Weinrobe (1983), who studied two alternative scenarios. Banks could either face extraordinary deposit withdrawals in the period following a natural disaster, an unexpected inflow of funds from payments by insurance claims, or aid transfer from government agencies. Generally, they do not find net outflows in post-disaster periods suggesting that scenario 2 seems to prevail. Recently, Allen et al. (2022) investigated whether small local banks (community banks) in the U.S. would potentially be exposed to liquidity risk after a major natural disaster. They find that in the quarter of the natural disaster, local banks increase lending, although deposits decrease. This pressure on a bank's balance sheet requires banks to access liquidity, which could occur through selling securities (stored liquidity) or through purchasing wholesale (brokered) deposits. Their results are consistent with local community banks using stored liquidity to respond to a natural disaster as an unexpected shock on both sides of their balance sheet. This corroborates prior research about the vital role that community banks place on stored liquid assets. They have historically self-managed their liquidity and avoided external funding in times of sudden shocks by selling part of their stored liquidity (DeYoung et al. 2018).

Bank exposure to liquidity risk is also investigated by Brei et al. (2019) who studied the adverse effects of hurricanes on a sample of East Caribbean banks. The impact of natural disasters could be more prolonged in developing markets since their financial systems are underdeveloped compared to the most advanced economies. Brei et al. (2019) show that following hurricanes, banks in the East Caribbean islands experienced massive withdrawals and a dry-up of non-deposit funding, which translated into a contraction of lending to households and businesses.

---

[4] U.S. banks pass through loans to a well-functioning securitization market. This contributes to an significant increase of post-disaster loans from all banks operating in stricken counties. More on this in the following sections.

Contrary to theoretical predictions, non-performing loans did not increase, nor did bank profitability and capital adequacy worsen. This outcome suggests a proactive strategy of central banks in small developing countries exposed to weather hazards. They should consider increasing reserve requirements during hurricane seasons and lowering them during normal times or in response to hurricane strikes so banks can build liquidity buffers and use them after the shock.

These findings are summarized in Fig. 3.2. The empirical literature so far shows that the most adverse effects of weather hazards are generally observed in less developed economies with weak institutional and regulatory frameworks. Catastrophic weather events impact loan quality, and bank liquidity when banks lack adequate liquidity buffers and extraordinary lines of credit from wholesale markets are missing. At the same time, banks have started incorporating disaster risk in their models and are tightening credit standards in more disaster-prone areas.

3.2   The Regulatory Framework on Bank Exposure to Disaster and Climate Risk

3.2.1   The European Context

The European Union has developed a strategy for promoting sustainable growth which dates back to two intergovernmental agreements, The United Nations 2030 agenda and the Paris Agreement on climate change. The latter sets out a global framework to avoid dangerous climate change by limiting global warming to well below 2 °C and pursuing efforts to limit it to 1.5 °C. It also aims to strengthen countries' ability to deal with the impacts of climate change and support them in their efforts toward achieving this target. The Paris Agreement is the first-ever universal, legally binding global climate change agreement, adopted at the Paris climate conference (COP21) in December 2015. The E.U. and its Member States are among the 190 Parties that signed the Paris Agreement. The E.U. formally ratified the Agreement on October 5, 2016, thus enabling its entry into force on November 4, 2016.

All European governments have agreed to fortify societies' capability to handle the consequences of climate change and offer sustained and improved worldwide assistance for adjustment to developing nations. The arrangement also acknowledges the significance of preventing, decreasing, and dealing with loss and damage linked with the unfavorable effects of

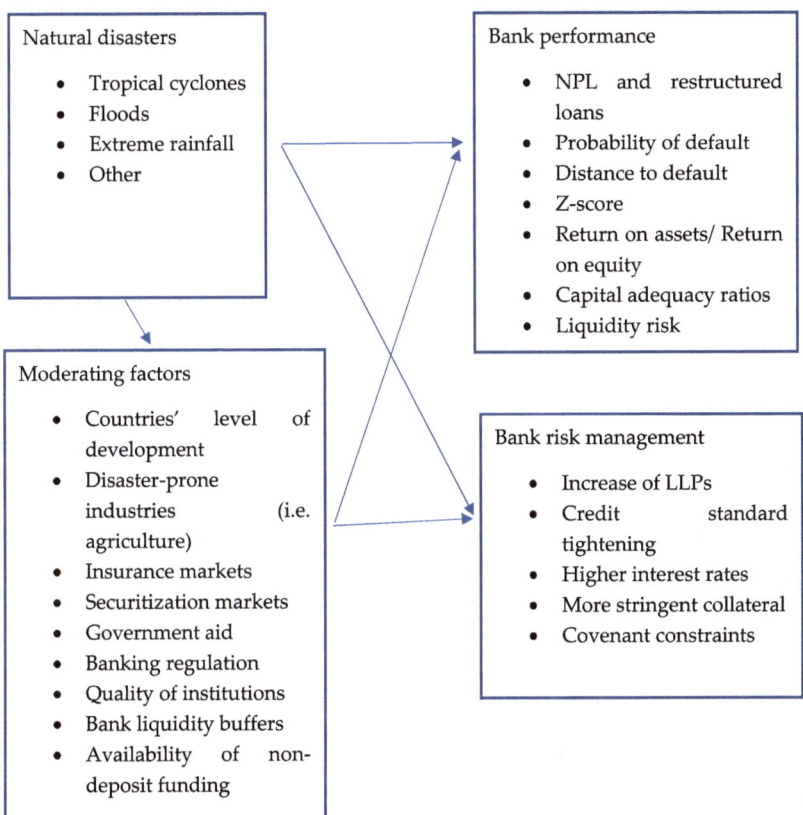

Fig. 3.2 Main findings of the empirical literature on the impact of natural disasters on bank risk and performance

climate change; acknowledges the necessity to collaborate and amplify the comprehension, deed, and support in various fields such as advance notice structures, emergency readiness, and hazard assurance.[5]

The European Commission published in 2018 an Action Plan for financing sustainable growth since the financial sector should play a key role in enabling the growth of investments that guarantee a financial

[5] https://climate.ec.europa.eu/eu-action/international-action-climate-change/climate-negotiations/paris-agreement_it.

return, but do not compromise environmental and social factors. At the same time, it is expected that financial institutions integrate sustainability into their risk management framework and promote transparency about their long-term investment strategies. In particular, banks should integrate climate risks into their credit risk models. Climate risks and exposure to them should be correctly measured through ESG ratings. Unfortunately, ESG ratings can diverge between different agencies due to methodological, data-related, subjective, and industry-specific factors. Investors and other stakeholders need to understand these differences and consider multiple ratings when making decisions (Berg et al. 2019).

The European Commission mandated the European Banking Authority (EBA) to assess banks' exposure to environmental and social risks by the end of 2025. The inclusion of ESG risks in the supervisory processes of regulatory authorities in the E.U. was validated by EBA in 2021. In December 2022, EBA published its road map outlining the objectives and the timeline for delivering mandates and tasks in sustainable finance and environmental, social, and governance (ESG) risks. The roadmap explains the EBA's sequenced and comprehensive approach over the next three years to integrate ESG risk considerations in the banking framework and support the E.U.'s efforts to achieve the transition to a more sustainable economy.

EBA will undertake numerous measures that have distinct objectives. Initially, it will guarantee that banks cultivate and implement a broader approach to divulge the assimilation of ESG risks in their processes. It will certify that ESG factors are sufficiently incorporated in institutions' risk management framework and into their supervision, including through additional advancements in climate stress tests. Concerning prudential regulation, the EBA has commenced an evaluation of whether modifications to the prevailing prudential treatment of exposures to encompass environmental and social deliberations would be warranted.[6]

More specifically, an analysis of large banks' exposure to physical and transition risks will be implemented. It aims to ensure that the risk management frameworks of institutions and the supervisory processes of competent authorities appropriately incorporate these risks. To pave the way for its strategy on climate risk stress tests, the EBA conducted in 2020 an EU-wide pilot exercise on climate risk and published the results in May

[6] https://www.eba.europa.eu/eba-publishes-its-roadmap-sustainable-finance.

2021. It was designed as a learning exercise where 29 banks volunteered to participate. This stress test aimed at measuring the impact of different economic scenarios on banks' stability, conditional on their exposure to climate risks, catastrophic weather events, and sustainability transition.[7]

3.2.2 The European Central Bank's (ECB) Action on Climate Risks

The European Central Bank (ECB) is taking several measures to address the climate risk exposure of E.U. banks. As part of its supervisory role, the ECB is currently assessing the extent to which banks are considering climate risks in their business operations and risk management. The ECB has also published guidance on how banks can effectively manage and disclose climate-related risks. Additionally, the ECB collaborates with other central banks and supervisory authorities to share knowledge and best practices on climate risk management. Finally, the ECB is conducting research to deepen its understanding of climate risks and their implications for financial stability.

More in detail, the ECB published in 2021 a report on the integration of climate risks in banks' internal stress tests.[8] From the report emerges that banks are focusing more on physical risks and losses related to natural disasters compared to transition risks, although research has shown the two types of risks are strongly correlated. The ECB is encouraging European banks to assess if their exposure to these risks eventually affects their regulatory capital ratios.

Starting in 2021, the ECB has commenced the implementation of annual climate stress tests on a large sample of European banks and their customers, projecting their activities over the next 30 years and hypothesizing different climate scenarios. These have been built by the Network for Greening the Financial System (NGFS),[9] and represent different possible climate conditions with a different impact of physical and transition risks on banks' profitability and resilience in the medium-long term.

---

[7] AIFIRM, Position Paper 39/2022.

[8] ECB (2021). The state of climate and environmental risk management in the banking sector.

[9] NGFS is a network of central banks and regulators whose purpose is to meet the goals of the Paris Agreement and to enhance the role of the financial system to manage risks and mobilize capital for green and low-carbon investments in the broader context of environmentally sustainable development.

Here, the results of the latest stress test conducted in 2022 will be briefly summarized.

Climate risks seem relevant for the large majority of significant institutions directly supervised by the ECB. All of them, to varying degrees, are exposed to the materialization of acute physical risks in Europe, especially drought, heat events, and flood risk. The risks that banks are facing in this regard are closely linked to the geographical location of their lending activities and could in some cases lead to non-negligible losses. For instance, banks whose portfolio is exposed to Southern Europe could face losses from wildfires or droughts. In contrast those institutions operating in Central and Northern Europe would be more exposed to floods. Firms located in these areas could suffer severe disruptions of their productive activities, which could increase their probability of default in the medium-long term due to the higher frequency of these events.

The stress test revealed that under a short-term, three-year disorderly transition risk scenario and two physical risk scenarios (flood risk and drought/heat risk), the combined credit and market risk losses for the 41 banks providing projections would amount to around €70 billion. Figure 3.3 describes more in detail the different scenarios.

There is a high probability that this test could underestimate potential losses for several reasons. First, the models did not incorporate any economic downturn accompanying the adverse climate effects. Second, the test was applied only to a subsample of the total exposure of the banks

	Expo sures	Scenario	Projections[1]	Horizon	Credit risk	Market risk	Operational risk
Transition risk	Global	Short-term stress — Baseline	3 years (2022-2024)	Corporate loans (incl. SME, CRE) + mortgages	Bonds + stocks issued by NFCs[2] (incl. accounting and economic hedges)	Operational and reputational risks to be assessed via a qualitative questionnaire
		Short-term stress — Stress					
		Long-term paths — Orderly	30 years (2030, 2040, 2050)	Corporate loans (incl. SME, CRE) + mortgages		
		Long-term paths — Disorderly				
		Long-term paths — Hot house				
Physical risk	EU countries	Drought & heat risk — Baseline	1 year (2022)	Corporate loans (incl. SME)	1.All projections with the exception of the long-term paths will be based on a static balance sheet.	
		Drought & heat risk — Stress			2.The parent company needs to be an NFC, e.g. bonds issued by car financing company X are in scope.	
		Flood risk — Baseline	1 year (2022)	Mortgages + CRE loans		
		Flood risk — Stress				

Fig. 3.3 ECB climate scenarios and risk dimensions (*Source* ECB climate risk stress test, methodology, October 2021)

analyzed. In Figs. 3.4–3.6, a graphical representation of the potential consequences of different physical risk scenarios for banks is shown.

In the drought and heat scenario, extreme temperatures do not affect sectors and countries uniformly. The impact mainly materializes for outdoor sectors such as agricultural activities, construction, or mining via decreased sectoral productivity (Fig. 3.4).

Loan losses predominantly derive from exposures to areas vulnerable to heat and drought, and the impact is heterogeneous across banks. However, the tests did not incorporate potential alleviating measures related to insurance and public aid schemes. Figure 3.5 evidences that loan losses will be much higher for exposures to countries that are more prone to heat and drought risks. At the same time, these countries will experience a more significant drop in the gross value added compared to the baseline scenario vis-à-vis other economies.

Flood risks, instead, impact on the value of the collateral. Real estate assets located in areas affected by floods should experience severe damage. Consequently, real estate prices would experience a negative shock, which should increase loan-to-value ratios, and ultimately would impact the

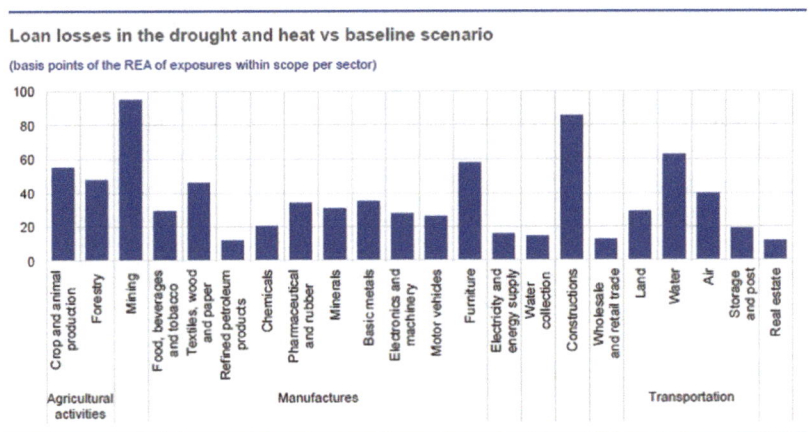

Fig. 3.4 Accumulated loan losses under the drought and heat scenario (*Source* ECB calculations)[10]

---

[10] REA stands for Risk Exposure Amount, which is the bank's exposure for a specific class of borrowers.

(basis points of the REA of exposures in scope (y-axis) vs cumulative GVA growth (x-axis))

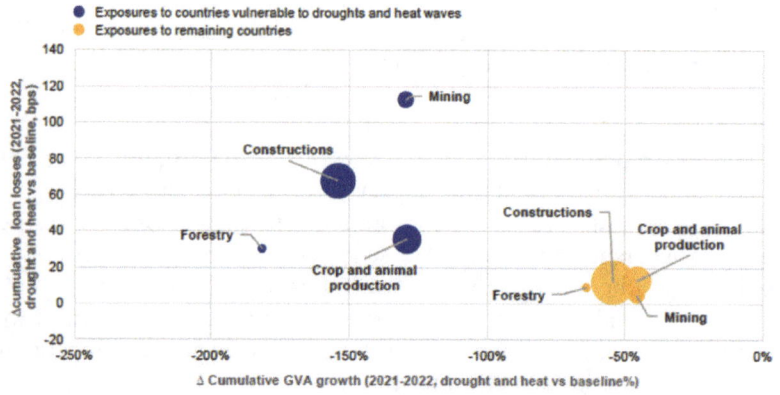

Fig. 3.5 Cumulative loan losses in drought and heat vs baseline scenario for vulnerable/non-vulnerable countries (*Source* ECB calculations)

expected losses through an increase of the loss given default (LGD). Figure 3.6 indicates that high and medium flood risk exposures bear half of the losses, with a share of just 31% of the exposures.

Other important stress test findings are that banks still lack a clearly defined long-term strategy for credit allocation policies that reflect transition to a more sustainable low-carbon reality. Around 60% of banks still have to adopt a well-integrated climate risk stress-testing framework, and most of those banks envisage a medium to long-term time frame for incorporating physical and/or transition climate risk into their framework.

The exercise also revealed that many banks are still at an early stage in factoring climate risk into their credit risk models. In many cases, credit risk parameters projected by banks were found to be quite insensitive to the climate risk shocks depicted in the scenarios.

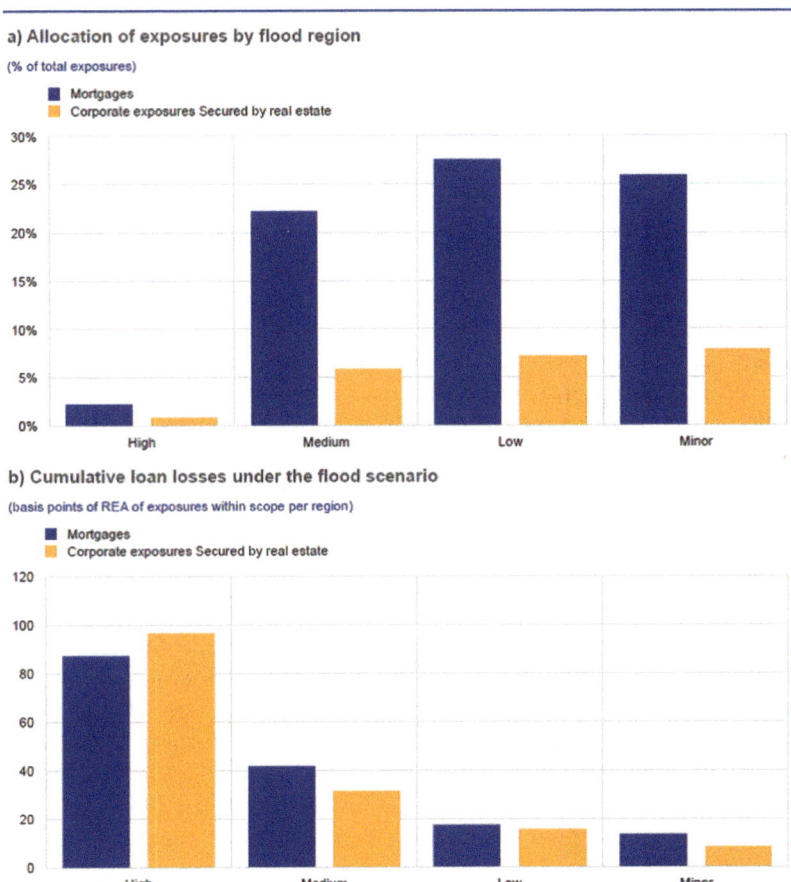

Fig. 3.6 Exposures and loan losses under the flood risk scenario (*Source* ECB calculations)

3.2.3   The U.S. Approach to the Impact of Climate Risks on Bank Stability

Climate risks are increasingly assuming importance for the FED policy. The most recent Financial Stability Reports[11] are including climate risk in the financial stability monitoring framework of the institution. The framework distinguishes between shocks to the financial system and the economy. The latter are difficult to predict, and vulnerabilities, defined as underlying features of an economic or financial system, can amplify the negative effects of shocks.[12]

Although the FED acknowledges that there is a broad consensus about the impact of climate change on the economy, indicates that there is a high level of uncertainty about the exact timing and precise magnitudes of future climate outcomes. Moreover, it outlines how several challenges related to measuring the clear impact of climate risks to the real economy and the financial sector remain. These arise because of the lack of granular data, unsophisticated mathematical models, and mismatching between mitigation costs which are incurred upfront and benefits which become more evident gradually over time.

Nevertheless, the FED started in 2022 to implement various steps aimed at assessing and promoting the resilience of the financial sector and in particular, of the largest banks under its supervision. The Federal Reserve Board is working closely with the Office of the Comptroller of the Currency (OCC) and the Federal Deposit Insurance Corporation (FDIC) to propose guidance for these institutions on identifying, measuring, monitoring, and management of climate-related financial risks. These principles will help large banks evaluate risk exposure and incorporate climate-related financial risks into their risk management frameworks.

In mid-2023, the FED Board is starting a pilot Climate Scenario Analysis (CSA) to learn about the six largest banking organizations' climate risk management practices and challenges and to enhance the ability of both large banking organizations and supervisors to identify, measure, monitor, and manage climate-related physical and transition financial risks.

---

[11] These are the documents through which the Federal Reserve Board summarizes the framework for assessing the resilience of the U.S. financial system.

[12] Brunetti et al. (2021). The text is available here https://www.federalreserve.gov/econres/notes/feds-notes/climate-change-and-financial-stability-20210319.html.

The FED has established in 2022 a Supervision Climate Committee, which brings together senior staff from the Federal Reserve Board and Reserve Banks across the System to ensure supervised firms are resilient to climate-related financial risks.

The Board's pilot CSA exercise will adopt the IPCC's illustrative greenhouse gases (GHG) concentration trajectories to evaluate how resilient are the participants' real estate credit portfolios to a range of physical risks events of varying severity.[13] These trajectories are considered plausible and illustrative scenarios, and do not have probabilities attached to them. They represent a widely referenced set of projections developed with input from domestic and international climate experts. Using a standard set of assumptions allows participants to focus on evaluating risks rather than developing the trajectories themselves.

3.2.4 *Other Stress Tests Conducted by Central Banks and Regulators Around the Globe*

Several regulatory authorities have implemented climate stress tests for banks and other intermediaries under their jurisdiction in recent years. These exercises are either based on a *bottom-up* or *top-down* philosophy. In the first case, national banks use their data and calculate their performance under different climate scenarios. In contrast, under the latter, the methodology and guidance is provided by the regulator. Table 3.1 presents the main climate stress tests performed in several advanced economies related to physical risks.

These stress tests in general, have sparked some criticism from academia for several reasons. First, many of them are in a pilot phase, with only a few concluded. Second, more sophisticated models are needed to capture how climate risks will likely materialize and impact the real economy in the medium-long term. One of the main areas for improvement is that models need to include feedback loops between banks and the real economy. Third, given the extremely high complexity of the variables tested, and the difficulty in assessing the impact of global warming on economic and financial systems, regulators should foster market actors

---

[13] The Intergovernmental Panel on Climate Change (IPCC) is an intergovernmental body of the United Nations charged with advancing scientific knowledge about anthropogenic climate change.

Table 3.1 Several physical risks climate stress tests run by national regulators in recent years (2020–2022)

Authority	Scenario	Type of intermediary	Focus
Prudential Regulation Authority (UK)	60 Years	Banks and large insurers	Sectorial exposures Mortgages Corporate exposure
Banque de France	30 Years	Banks and Insurers	Business loans Equity corporate credit spreads Sovereign credit spreads Commodities interest rate
Australian Prudential Regulation Authority	30 Years	Banks	Mortgages and businesses
Hong Kong Monetary Authority	5–30 Years	Banks	Sectoral exposures

toward a managed transition to minimize harm to the financial system and the economy rather than engaging in highly unrealistic climate tests.

3.2.5 Why Central Banks Should Include Climate Risks in Their Mandates (and Why Not)?

Natural disasters as a consequence of climate change and global warming bear substantial effects both on the demand and supply side of the economy, as Table 3.2 evidences. Demand-side shocks are related to private and government consumption and investment, business investment, and international trade (Bolton et al. 2020a). This means that even non-exposed countries will be affected due to interconnection through global trade.

On the other side, as previously mentioned, natural disasters affect all potential supply, labor, technology, and physical capital components. Workers' productivity drops due to extreme temperatures, wages, and labor markets that are affected by a displacement of populations to other regions. Capital is diverted toward investments to contain climate damage, and its depreciation rate is faster (Opitz Sapleton et al. 2017; Bolton et al. 2020b).

Table 3.2 Climate change shocks and their impact on global demand and supply

	Shock	Global warming	Extreme weather (Natural disasters examples)
Demand	Investment	Uncertainty about future demand and climate risks	Uncertainty about climate risk
	Consumption	Changes in consumption patterns	Increased risk of losses from hurricanes or tropical cyclones
	Trade	Changes in transport systems	Disruption of supply chains
Supply	Labor	Loss of hours due to increased temperatures	Loss of hours due to heatwaves
	Capital stock	Diversion of resources toward adaptation capital	Damage due to extreme weather
	Technology	Loss of productivity	Diversion of resources from innovation to reconstruction and replacement

Source Adapted from NGFS (2019) and Bolton et al. (2020a)

The impact of disaster risks and climate events on price stability has yet to be examined accurately given its extreme complexity. There could be temporary or permanent pressures on inflation after certain natural disasters. For example, food prices could increase. In addition, agricultural production in certain countries could be adversely affected permanently. Conversely, there could be demand shocks due to loss of wealth which could pull inflation in the opposite direction, and depress economic growth. Central banks could face a dilemma between their inflationary target objective and reviving economic growth (Batten 2018; Debelle 2019).

Olovsson (2018) evidences that climate change has long-lasting stagflationary effects, which central banks cannot successfully confront. A coordinated response from many countries would be necessary, whereas actions taken by a single regulator could have scarce success.

While the link between climate risks such as natural disasters and monetary policy looks complex and advocates further research, the former's potential impact on the financial sector's stability has been widely recognized. This justifies the attention that central banks and other regulators are dedicating to the problem, as evidenced in the previous section. As they increase worldwide, natural catastrophes severely threaten the financial solvency of households, businesses, governments, and consequently banks and insurance companies, especially in poor and middle-income countries. Insured losses, on their end, may place insurers and reinsurers in a situation of fragility as claims for damages keep increasing.

In addition to these considerations, since the last Global Financial Crisis, central banks have increasingly pledged to enforce the stability of financial institutions by curtailing systemic risks that banks pose to the economy. Catastrophic natural disasters could present a source of systemic risk to the financial sector. Therefore, from a financial stability perspective, central banks must include natural disaster risk in their prudential policy frameworks (Bolton et al. 2021). However, prominent scholars have expressed their skepticism about the ability of central banks to successfully deal with natural disasters and climate change. Jean Tirole, for instance, has warned that although climate risks should be included in the risk management framework of banks and their regulators, the fight against climate change is not a task of the central banks but up to society as a whole. John Cochrane rejects the idea of central banks' mandates incorporating climate risks. Instead, they should focus on something they have a long experience with, such as inflation targeting. Jens Weidmann, president of the Deutsche Bundesbank, has marked out that discriminating against polluters is not the task of the financial sector in Europe but pertains to the activities of the Governments and other authorities. The Chairman of the FED, Jerome Powell, recently asserted that climate change is a matter of interest for the FED but cannot be considered its top priority.

It is reasonable to assume that central banks can risk losing their legitimacy if overly broad powers are delegated to such unelected independent institutions (Bolton et al. 2019). Central banks do not have the magic potion to solve the climate crisis on their own, even if they possess powerful tools at their disposal that could be applied to reverse carbon emissions. However, equally central banks must be part of the solution, and must play their part along with other actors in the drive to fight climate change in our economies.

3.3   How Could Banks Integrate Climate Risks into Their Operations?

Physical risks generate various consequences on firms' business strategy, product mix, the geographic distribution of the activities, type of business model, operations, etc. For instance, corporations might consider positioning their operations and manufacturing facilities based on the exposure to certain physical risks. Consequently, firm performance will be affected by increasing adaptation costs, alteration of revenue drivers (if the product mix will be modified), and operating costs (because of changes in productivity, for example).

Banks' asset allocation will be affected by the choices undertaken by their borrowers, and critical drivers that reflect their adaptation and response to climate risks. Banks and their corporate borrowers will be scrutinized by their stakeholders and both will need to improve the disclosure of plans to identify and mitigate adverse financial impacts of climate change, transition to a net-zero economy, and how they intend to capture associated opportunities with this scenario.

In this regard, a recent report by the Task Force on Climate-Related Financial Disclosure (TFCD), created by the Financial Stability Board, recommends that organizations provide detailed disclosure around 4 thematic areas related to climate-related risks. These areas are related to core elements of how organizations operate: governance, strategy, risk management, metrics, and targets (Fig. 3.7).

In general, the framework for environmental risk analysis and management involves four steps (NGFS 2020)[14]:

- Risk identification

Exposure mapping and risk measurement methodologies for climate-related financial risks can be differentiated according to physical risk drivers. Banks should conduct strategic assessments of environmental factors that may cause financial risks (e.g., value impairment from extreme weather events such as floods, hurricanes, or droughts, devaluation of

---

[14] As previously mentioned, although this framework regards physical and transition risks, the main focus will be physical ones.

Governance
The organization's governance around climate-related risks and opportunities

Strategy
The actual and potential impacts of climate-related risks and opportunities on the organization's businesses, strategy, and financial planning

Risk Management
The processes used by the organization to identify, assess, and manage climate-related risks

Metrics and Targets
The metrics and targets used to assess and manage relevant climate-related risks and opportunities

Fig. 3.7 A voluntary framework of recommendations that incorporate climate-related risk considerations into financial disclosures (*Source* TCFD)

associated infrastructure, interruption of supply chains, and increased natural capital costs).

- Risk exposure

The Basel Committee has recently published a guideline on how banks can measure the impact of physical risk (and transition risk) on their overall risk exposure.[15] Physical risk drivers (physical hazards) can be linked to financial exposures using damage functions that define the impacts of specific hazards on the tangible assets and activities that generate financial flows. The disruptions to assets, activities, and their corresponding financial flows can then be integrated, in principle, into established risk models that dimension financial risk parameters. Additionally, specific damage functions will be informed by sectors, the severity of hazards, time horizon factors, and geospatial idiosyncrasies.

- Risk assessment

An estimation of the probabilities and magnitudes of financial losses arising from these risks (using scenario analysis and stress tests) is necessary. The results of these models could feed into risk pricing.

[15] BCBS (2021).

Banks are exposed to climate-related financial risks through their transactions with clients, and counterparties: corporates, households, sovereigns, and other financial institutions. Banks and supervisors should determine the level of exposure granularity most relevant for their respective risk assessments, when estimating the implications of climate risk drivers for these transactions. This may be influenced by several factors, such as specific physical risk drivers, availability of relevant data, risk management decision-making processes, and increasing computational complexity with increasing granularity.

Banks can choose between a *top-down* or *bottom-up* approach. In general, *top-down* approaches start by dimensioning risk at the general, or aggregated, level and then "push down" or attribute the aggregated measure of risk to the single counterparties.[16] *Bottom-up* modeling approaches usually begin by assessing the impact of physical and/or transition risks on an asset, counterparty, or portfolio level, enabling differentiating of financial impacts based on specific exposure characteristics and then aggregate up to a consolidated measure of (i.e., credit or market) risk. Aggregating individual risk exposures to a consolidated view of risk may require understanding potential correlations among exposures that could amplify or diversify risk within a portfolio or bank.

- Risk mitigation

As banks consider how to measure climate-related financial risks, they may need to estimate the effect of potential risk mitigation and to what extent mitigants could moderate or offset risk-taking. In addition to proactive measures a bank can take to reduce its risk exposure, climate-related financial risks can also be offset through counterparty measures to adapt to or mitigate the effects of climate change. These measures may alter the relationship of exposures to risk drivers within banks' risk models. For example, strong building regulations and flood mitigation measures may minimize an area's vulnerability to specific physical hazards, reducing potential losses on an asset exposed to the location, compared to an area without such regulations and measures (NGFS 2020b).

Commercial banks are starting to assess the impact of environmental risks on their traditional risk metrics, such as probabilities of default (PD),

---

[16] AIFIRM Position Paper 39/2022.

and loss given default (LGD). In a nutshell, these models first estimate losses as a consequence of specific metrics that are explanatory variables for the loan-related risk models (NGFS 2020b). These inputs are then utilized to estimate climate-adjusted PDs, LGDs, and credit ratings. An example of these models is the one developed by Tsinghua University in China which should support banks in analyzing credit risk arising from the impact of physical risks under various climate scenarios. This framework could be used for a wide range of hazards, including typhoons, floods, heat waves, drought, etc. It can be applied to a large variety of sectors, especially those vulnerable to natural disasters, for instance, housing, agriculture, energy, and transportation sectors.[17]

Although climate risk ratings or scores could be used as proxies of climate-related factors when granting credit, banks have integrated these scores into their overall customer credit rating only on targeted occasions so far. A relatively few banks is gradually integrating climate-related ratings into their counterparty credit assessment processes and risk management frameworks (NGFS 2020d). An example of such a climate risk assessment adopts a traffic light classification (red/amber/green) for differentiating clients according to their relative exposure to climate-related risks, where clients exposed to elevated climate-related risk are assigned a "red" or "amber" rating. This rating may justify a heightened credit review, monitoring, or approval protocols (ECB 2020).

In July 2018, 16 leading banks convened by the UNEP FI and supported by the consultancy Acclimatise released a methodological framework on physical climate risk analysis (UNEP FI 2018). The framework seeks to help banks make in-house estimates of the impact of physical climate risks on their loan portfolios, expressed by key credit risk metrics such as PDs and LTVs. The methodological framework is explicitly piloted for agriculture, energy, and real estate portfolios.[18]

Other models have the scope to support banks in estimating the impact of environmental risks on the performance of bonds, equity, or other securities held in the portfolio (i.e., market risk). These models first estimate the changes induced by environmental risks or factors to specific metrics that later constitute the determinant variables of these assets' valuation models.

---

[17] Chapter 6 of NGFS (2020c).
[18] UNEP FI (2018) and Chapter 13 of NGFS (2020c).

Recently, Natural Capital Finance Alliance (NCFA) which is a collaboration between UNEP FI and Global Canopy, developed a process that enables financial institutions to assess natural capital risks in their portfolios easily, outlined in the report "Integrating Natural Capital in Risk Assessments".[19] The step-by-step guide for conducting a rapid natural capital risk assessment focuses on identifying how businesses depend on the environment, how these dependencies are threatened by environmental change, and the resulting risks for financial institutions. The guide, released in January 2019, assists global banks in better understanding how the pollution of oceans or the destruction of forests, for example, may affect their financial future.

3.3.1 The Criticality of Data Granularity and Models' Weaknesses

Data on climate risks and natural disasters are usually provided by government agencies and academic institutions. The datasets include forecasts for fluvial and coastal flooding or wildfires and in some cases, provide probability distributions for the hazard event. In addition, physical risk measurement approaches may also include the vulnerability of a location to physical hazards, which requires combining information on past hazards or forward-looking hazard projections with information on location-specific characteristics such as land, elevation, soil composition, or land cover (BCBS 2021). Consequently, to assess exposures' vulnerability to physical risks, a comprehensive matching of physical hazards with the location of relevant physical touchpoints can facilitate the estimation of potential economic impacts and, ultimately, the potential disruption to the contractual cash flows of a bank's lending exposure.

One bank in the Netherlands stressed its residential mortgage portfolio to assess its exposure to flood risks (Netherlands Bank 2020). High-risk regions were identified using publicly available granular flood risk maps (i.e., postal code level). The bank performed a geospatial mapping of its portfolio on these high-risk regions to identify exposure to elevated flood risk.

However, more than knowing the exact location of the firm's headquarter may be required to assess the risk of a shock to its full value chain arising from physical hazards. Additionally, information on the

---

[19] The report is available on this weblink. https://www.unepfi.org/publications/integrating-natural-capital-in-risk-assessments/.

interconnectedness of counterparties (i.e., corporates, municipalities) to surrounding regions and demand markets may be needed to assess the impact of physical risk events on revenue forecasts or labor productivity.

Most physical risk analyses focus on direct physical damages on properties, infrastructure, and agriculture assets, with limited reference to the impact of climate events on variables affecting firms' operating environment. For example, it has been challenging for these analyses to quantify the relationships between natural disasters and the resulting damages to local economic growth, household income, unemployment rate, and supply chain conditions. In addition, models often ignore the macroeconomic feedback loops despite the fact that environmental and climate changes may well impact many macroeconomic and macro-financial variables that will drive company performance.

REFERENCES

Albuquerque, P.H., Rajhi, W., 2019. Banking stability, natural disasters, and state fragility: Panel VAR evidence from developing countries. Research in International Business and Finance, 50, pp. 430–443.

Allen, K.D., Whitledge, M.D., Winters, D.B., 2022. Community bank liquidity: Natural disasters as a natural experiment. Journal of Financial Stability, 60, p. 101002.

Anderlik, J., Bush, A., Cofer, R.J., Kalser, K., McGee, J., 2022. Severe weather events and local economic and banking conditions. Federal Deposit Insurance Corporation Staff Studies, Report No. 2022-03.

Apergis, N., 2022. Do weather disasters affect banks' systemic risks? Two channels that confirm it. Economics Letters. https://doi.org/10.1080/135 04851.2022.2084015.

Batten, S., 2018. Climate change and the macro-economy: A critical review. Bank of England Working Paper No. 706.

Berg, F., Koelbel, J., Rigobon, R., 2019. Aggregate confusion: The divergence of ESG ratings. MIT Working Paper 2019.

Berger, A.N., Molyneux, P., Wilson, J.O.S., 2019. Banking: A decade on from the global financial crisis. In Allen N. Berger, Philip Molyneux, and John O. S. Wilson (eds.), The Oxford Handbook of Banking, 3rd edn. Oxford Handbooks, Oxford Academic.

BCBS, 2021. Climate-related risk drivers and their transmission channels. www.bis.org/bcbs/publ/d517.pdf.

Blickle, K.S., Hamerling, S.N., Morgan, D.P., 2022. How bad are weather disasters for banks? Federal Reserve Bank of New York Staff Reports, No. 990.

Bolton, P., Cecchetti, S., Danthine, J.P., Vives, X., 2019. Sound at last? Assessing a decade of financial regulation. Centre for Economic and Policy Research, London, UK.

Bolton, P., Despres, M., Pereira da Silva, L., Samama, F., Svartzman, R., 2020a. The green swan: Central banking and financial stability in the age of climate change. Bank of International Settlements and Banque de France.

Bolton, P., Despres, M., Pereira da Silva, L., Samama, F., Svartzman, R., 2020b. Green swans: Central banks in the age of climate related risks. Banque de France Bulletin 229.

Bolton, P., Kacperczyk, M., Hong, H., Vives, X., 2021. Resilience of the financial system to natural Disasters. Centre for Economic and Policy Research, London, UK.

Brei, M., Mohan, P., Strobl, E., 2019. The impact of natural disasters on the banking sector: Evidence fromhurricane strikes in the Caribbean. The Quarterly Review of Economics and Finance, 72, pp. 232–239.

Brunetti, C., Dennis, B., Gates, D., Hancock, D., Ignell, D., Kiser, E.K., Kotta, G., Kovner, A., Rosen, R.J., Tabor, N.K., 2021. Climate change and financial stability. Federal Reserve Board of Governors FEDS Notes, March 19.

Collier, B., Katchova, A.L., Skees, J.L., 2011. Loan portfolio performance and El Niño, an intervention analysis. Agriculture Finance Review, 71, pp. 98–119.

Dal Maso, L., Kanagaretnam, K., Lobo, G.J., Mazzi, F., 2022. Does disaster risk relate to banks' loan loss provisions? European Accounting Review, forthcoming.

Debelle, G., 2019. Climate change and the economy. Speech at the Public Forum hosted by the Centre for Policy Development, Sydney, 12 March 2019. Available at: https://www.bis.org/review/r190313d.pdf.

DeYoung, R., Distinguin, I., Tarazi, A., 2018. The joint regulation of bank liquidity and bank capital. Journal of Financial Intermediation, 34, pp. 32–46.

Duanmu, J., Li, Y., Lin, M., Tahsin, S., 2022. Natural disaster risk and residential mortgage lending standards. Journal of Real Estate Research, 44, pp. 106–130.

ECB, 2020. Guide on climate-related and environmental risks: Supervisory expectations relating to risk management and disclosure.

ECB, 2021. The state of climate and environmental risk management in the banking sector.

He Huang, H., Kerstein, J., Wang, C., Wu, F., 2022. Firm climate risk, risk management, and bank loan financing. Strategic Management Journal, 43, pp. 2849–2880.

Klomp, J., 2014. Financial fragility and natural disasters: An empirical analysis. Journal of Financial Stability, 13, pp. 180–192.

Kundu, S., Park, S., Vats, N., 2022. The geography of bank deposits and the origins of aggregate fluctuations. Available at SSRN: https://ssrn.com/abstract=3883605

Netherlands Bank, 2020. Good practice—Integration of climate-related risk considerations into banks' risk management. https://www.toezicht.dnb.nl/2/50-238193.jsp.

NGFS, 2019. Macroeconomic and financial stability: Implications of climate change. NGFS Technical Supplement to the First Comprehensive Report. Available at: https://www.ngfs.net/en/first-comprehensive-report-call-action.

NGFS, 2020a. NGFS climate scenarios for central banks and supervisors. www.ngfs.net/sites/default/files/medias/documents/820184_ngfs_scenarios_final_version_v6.pdf.

NGFS, 2020b. Overview of environmental risk analysis by financial institutions. https://www.ngfs.net/en/overview-environmental-risk-analysis-financial-institutions.

NGFS, 2020c. Case studies of ERA methodologies. NGFS Occasional Papers. https://www.ngfs.net/en/case-studies-environmental-risk-analysis-methodologies.

NGFS, 2020d. A status report on Financial Institutions' experiences working with green, non-green and brown financial assets and a potential risk differential. https://www.ngfs.net/sites/default/files/medias/documents/ngfs_status_report.pdf.

Noth, F., Schüwer, U., 2023. Natural disasters and bank stability: Evidence from the U.S. financial system. Journal of Environmental Economics and Management, 119, p. 102792.

Olovsson, C., 2018. Is climate change relevant for Central Banks? Sveriges Riksbank Economic Commentaries. https://doi.org/10.2307/2551631.

Opitz Sapleton, S., Nadin, R., Watson, C., Kellett, J., 2017. Climate change, migration, and displacement: The need for a risk-informed and coherent approach. The Encyclopedia of Global Human Migration. https://doi.org/10.1002/9781444351071.

Steindl, F.G., Weinrobe, M.D., 1983. Natural hazards and deposit behavior at financial institutions. Journal of Banking and Finance, 7, pp. 111–118.

UNEP FI, 2018. Navigating a new climate: Assessing credit risk and opportunity in a changing climate: Outputs of a working group of 16 banks piloting the TCFD recommendations PART 2: Physical risks and opportunities, UNEP Financial Initiative & Acclimatise. Geneva, Switzerland.

Von Dahlen, S., Von Peter, G., 2012. Natural catastrophes and global reinsurance—Exploring the linkages. BIS Quarterly Review, December 2012, 23–35.

CHAPTER 4

The Role of Banks in Promoting Post-disaster Economic Growth

Abstract Academic research on post-disaster recovery acknowledges that growth stems from increased bank lending to accommodate reconstruction efforts. Local banks are more active than multimarket ones in addressing the increased credit demand because their relationship lending approach allows them to evaluate borrowers' likelihood of default correctly. At the same time, other factors matter, such as bank capitalization, securitization markets, and government subsidies. Banks' lending strategies can no longer avoid the impact of increased climate risk on their future performance.

Keywords Bank activity and economic growth · Bank deregulation · Financial development · Post-disaster lending · Local bank

© The Author(s), under exclusive license to Springer Nature Switzerland AG 2023
A. Duqi, *Banking Institutions and Natural Disasters*,
Palgrave Studies in Impact Finance,
https://doi.org/10.1007/978-3-031-36371-9_4

4.1   The Theoretical Framework Underlying the Link Between Bank Activity and the Real Economy

Banks are vital for every economy because they offer payment services that facilitate trade, channel savings from investors to firms, households, and governments, diversify and reduce liquidity and intertemporal risk (Beck 2008). Economic theories claim that banks accumulate an advantage over time in collecting and processing information, enabling them to reduce adverse selection and moral hazard and allocate money to those that can make better use of it (Bencivenga and Smith 1991; Levine et al. 2000). By providing these services to the economy, financial intermediaries influence savings and allocation decisions that may alter long-run growth rates.

However, the nexus of causality between the level of development of the banking sector (or overall financial sector) in a country and real economic activity is still a debatable topic in academia for reasons that will be outlined shortly. This debate has stimulated research that has initially benefitted from detailed country-level data and then has increased in granularity by offering an industry and firm-level analysis. Therefore, this section aims to present the main findings of this literature and add further insights into the links between banks, the real economy, and the mechanisms through which finance might enhance growth.

One of the main concerns when linking the financial sector with the real economy is how to measure a well-functioning financial environment. Over time several variables have been used, such as the ratio of liquid liabilities of intermediaries operating in a country divided by its GDP, the ratio of deposits of domestic banks relative to the central bank's total assets, credit issued to non-financial firms divided by total credit to the economy or divided by countries' GDP. It can be noticed that all these variables are primarily related to the banking sector in each country, especially in emerging economies.

The first contributions to this field are King and Levine (1993) and Levine and Zervos (1998), who find that higher levels of financial development are positively associated with faster rates of economic growth, physical capital accumulation, and economic efficiency improvements both before and after controlling for numerous country and policy characteristics. Furthermore, higher levels of financial development are strongly associated with future capital accumulation rates and improvements in the efficiency with which economies employ capital. The authors link these results with the famous work of Schumpeter (1911), who stated that bank credit does create real value for society.

However, these first contributions suffer from several pitfalls. The models do not rule out that the development of the financial sector is clearly endogenous and could be influenced by the country's economic growth, which generates an estimation bias from simultaneity, reverse causality, or omitted variables. Both the level of financial development and economic growth could be influenced by other missing variables. Another primary concern regarding these papers is that the variables that proxy for the banking activity are somehow a consequence of it and maybe change together with the economy. Therefore, they do not clearly tell us how banks generate an impact on society (Cetorelli and Blank 2019). Moreover, banks are in competition with stock markets in providing corporate finance. Beck and Levine (2004) find that the level of development of both is essential for economic growth.

4.1.1   The Importance of the Institutional and Legal Environment

In the following contributions, several authors have attempted to use more sophisticated models, such as GMM or Instrumental Variable Regression, to account for the abovementioned biases. In the first case, the lagged values of the dependent variable, which could be the country's GDP per capita real growth rate, are used as instruments of the current level of financial development. Beck et al. (2000), using a GMM specification, confirm the positive link of financial development with growth by improving countries' total productivity. Still, the impact on capital growth and savings rates remains ambiguous. Levine and Warusawitharana (2021) reach similar conclusions by focusing on the impact of financial markets' imperfections on the productivity growth of private European firms.

In the second case, the level of financial development is instrumented by an exogenous variable, namely, the country's legal system origin (English, French, German, or Scandinavian). This choice draws from the seminal paper of La Porta et al. (1998), who find that the current level of the regulatory and legal environment in many countries originates back to these four types inherited from a colonialist or occupation past. Since the essential activity of financial markets counts on the possibility of writing well-defined contracts regarding transactions based on promises of future payments, financial markets will be more or less developed to the extent to which the legal system allows the protection and enforcement of such contracts. Furthermore, since the establishment of a given legal system in a country is, to a large extent, the result of past events, such as experiences of colonization, it is plausible to consider this feature as exogenously determined (Cetorelli and Blank 2019).

From La Porta et al. (1998), a large body of literature has grown, linking countries' legal and regulatory frameworks and economic development. Levine (1998, 1999) examine the link between solid protection of creditor rights in different countries, degree of enforcement of contracts, development of the banking sector, and growth. In all specifications, these variables explain, to a large extent, that a strong banking sector promotes higher economic growth through physical capital accumulation and productivity improvements. Levine et al. (2000) instrument the level of financial development with countries' legal origins and reach the same conclusions as the previous studies corroborating the view that a higher level of financial development fosters growth.

Beck et al. (2003) confirm that the level of financial development in former colonies depends on the legal environment that colonialists created when they established their presence there.[1] Claessens and Laeven (2003) evidence that firms can grow in countries with substantial property rights because their intangible assets (patents, trademarks) cannot be extracted easily by competitors. More recently, Hasan et al. (2009) found that even in China, a strongly authoritarian economy, the empirical evidence suggests that the development of financial markets, legal environment, awareness of property rights, and a certain degree of political pluralism are associated with more robust regional growth. Demirgüç-Kunt et al. (2017) show that better corporate disclosure, indicated by adopting higher accounting standards, mitigates the destabilizing influence of limited access to long-term finance. Changes in financial development that tend to lengthen the maturity of credit have a potentially beneficial economic effect in terms of lower economic volatility, independently of the overall availability of external finance.

From this stream of research, it seems that various underlying forces (political, economic and financial conditions) are at play which might affect the link between countries' level of development of the financial

---

[1] According to Beck et al. (2003), countries' institutions were shaped by: the legal origin of its colonizers and the type of colonization strategies. These were aimed at favoring institutions to support private property rights in those countries where Europeans settled or to support the extraction of resources such as gold, silver, etc.

sector and their growth path. Several papers have added to this body of research by highlighting the critical role of the institutions' quality and stability and the government's accountability in each country (Beccera et al. 2012). Campos et al. (2012) confirm that political instability directly affects the impact of financial development on growth, as documented by the Argentinian economic decline over the twentieth century.

4.1.2   Does the Type of Bank Shareholder Matter?

Bank ownership has attracted interest since, in many countries, financial institutions are controlled either by governments or by important blockholders which are strongly connected to the political constituencies in those countries. Regarding the first type of owner, governments have aimed to control banks for "development" purposes (Gerschenkron 1962). Public banks were widespread until the 90s, even in developed economies such as Germany, France, Japan, and Italy. They are still operating nowadays across numerous emerging countries. The economic theory has advanced two alternative theoretical views to justify government ownership of banks. On the one side, development economics advocated state-owned banks along with the control of other strategic industries to foster their growth, especially when private capital is scarce, and the level of institutions is weak.

On the other hand, the "political" view of government participation in finance shares with the development view the desire of politicians to control local corporates' investment but emphasizes political rather than social objectives. According to this view, governments acquire control of enterprises and banks to provide employment, subsidies, and other benefits to supporters, who return the favor in the form of votes, political contributions, and bribes (Shleifer and Vishny 1994).

Is government ownership of banks beneficial for countries' growth? La Porta et al. (2002) find that the former is associated with slower subsequent development of the financial system, lower economic growth, and, in particular, lower productivity growth. Similar outcomes are found by Sapienza (2004) for Italy, Khwaja and Mian (2005) for Pakistan, and Imai (2009) for Japan.

However, other contributions suggest a more nuanced view. When the level of institutions in a country is high, state-owned banks can play a beneficial role in growth (Körner and Schnabel 2009; Andrianova et al. 2012; Önder and Özyıldırım 2013). State-owned banks allocate more credit to less advantaged regions, especially after a crisis, acting like a crisis resolution vehicle. Private banks in this setting tend to overreact to recessions and amplify the economic shock by reducing credit to local firms and households (Önder and Özyıldırım 2013).

Another related stream of literature looks at the "capture" of financial institutions by elites or powerful interest groups. This body of research links to the seminal work of Rajan and Zingales (2003), who evidence how the persistent financial underdevelopment of some countries and the full-scale reversal of financial development in others posit the "elite capture" of local financial systems. These contributions find that banking systems do not promote efficient capital allocation when wealthy tycoons or business families predominantly control them (Acemoglu et al. 2005; Fogel et al. 2008; Morck et al. 2011). Powerful families can extract rents from banks they own, use them to limit capital to potential competitors, or magnify agency problems when banks are part of large conglomerates, such as in several Asian and Latin American countries.

Another body of research has studied the differences between local and foreign banks and their impact on economic output. On the one hand, foreign banks can supply additional loanable funds and might have an advantage in overcoming informational and legal obstacles to lending, which benefit real economic activity (Giannetti and Ongena 2009; Bruno and Hauswald 2014). On the other hand, their activities might also exert competitive pressures on the local banking industry, which cuts back its lending activities, thereby hurting the overall provision of credit to firms and growth (Gormley 2008; Detragiache et al. 2008).

4.1.3   External Financial Dependence, Bank Market Power, and Growth

The endogeneity issue of financial development and the bias from omitted variables has been at the center of another important paper that deals with finance and growth. Rajan and Zingales (1998) investigate an alternative and overlooked mechanism behind the positive impact of financial development on economic output. They posit that financial development could help industries more in need of external finance, therefore offering support for intermediaries' role in reducing information asymmetries in financial markets.

The deregulation of the banking sector that started in the U.S. in the 1970s provided researchers with a natural experiment to test the link between banking and growth. It allowed them to disentangle better the contribution of the banking sector from other confounding effects. Until then, the U.S. had a fragmented banking market since interstate banking was not allowed. The end of the 1970s marks the beginning of an intense deregulation process, in which individual states—at different points in time—removed regulatory barriers that had prevented out-of-state bank entry. By the mid-1990s, the process had concluded, allowing from that point banks originally headquartered anywhere to expand potentially anywhere else without restriction. The banking market in the U.S. became, therefore, more competitive and efficient. In a seminal paper, Jayaratne and Strahan (1996) find that removing entry barriers to nonlocal banks benefitted state income growth. This result is much stronger from an econometrical point of view compared to previous studies that looked at proxies of financial development since it is related to the effect of a specific event, bank deregulation, that is supposed to spur growth. This occurs because, as theory predicts, more competitive banking sectors should allocate capital more efficiently.

After Jayaratne and Strahan (1996), numerous contributions start looking more closely at the link between bank market structure, external financial dependence, availability of credit, and firm growth. On these grounds, Cetorelli and Gambera (2001) adopt a cross-country approach to study the impact of bank concentration on economic growth and find that it imposes a deadweight loss that depresses growth. Nevertheless, they also show that the link between credit concentration and growth is heterogeneous across different industrial sectors, given that specific industries benefit from an environment of low bank competition. These industries are more in need of external finance, as Rajan and Zingales (1998) suggested. However, Claessens and Laeven (2005) reach opposite results insofar they show that industries most dependent on bank financing grow less in the presence of low bank competition. In another setting, the E.U. harmonization of banking, which essentially deregulated cross-border banking inside the EU, Romero-Ávila (2007) found that it significantly contributed to higher output growth through the improvement of the efficiency of financial intermediation by permitting a more productive use of resources and therefore decreasing the cost of capital for firms.

The above inconclusive results could be a consequence of the fact that there are two theoretically opposite views linking bank competition with better or worsened access to credit. On the one side, there is the conventional wisdom that more competition equals more credit since banks are encouraged to attract customers and increase market share. On the other hand, Petersen and Rajan (1995) prove that banks need at least some degree of market power to have the right incentives to undertake the proper investments in screening and monitoring necessary to resolve uncertainty about the quality of new entrepreneurs. The intuition is that in the absence of some ability to "capture" the client over time, a bank anticipates that an entrepreneur who turns out to be successful can seek better terms from competing banks that would not need to incur any upfront cost of screening and monitoring (or would spend just a fraction of what the original bank had to). A bank with market power could instead offer better initial terms knowing that any initial cost in starting such a lending relationship could be recovered later. The unconventional prediction that follows is that especially young and more opaque firms have better access to credit if they operate in more concentrated banking markets. Several studies have confirmed that this view holds under certain conditions. They find that firms belonging to more informationally opaque sectors grow more in the presence of concentrated banking markets (Bonaccorsi di Patti and Dell'Ariccia 2004; Inklaar et al. 2015).

Studies that test the competing theoretical views about the impact of bank market structure on growth have also looked at the creation and survival rate of new firms after banking markets were deregulated in the U.S. Even in this case, the results are not conclusive. On the one side, Black and Strahan (2002), Cetorelli (2003), and Cetorelli and Strahan (2006) document that after deregulation, the number of businesses increased. Moreover, the persistence of younger businesses was higher after deregulation. Cetorelli and Strahan (2006) evidence that more vigorous competition leads to more firms in operation and that the average firm size decreases as banking markets become more competitive. Smaller average firm size is consistent with the finding of more firms in operation, and both add strength to the idea that more bank competition favors and allows entry at a smaller scale.

Acharaya et al. (2011) confirm that deregulation increased growth output in those states and industrial sectors characterized by small and young firms, which rely on external finance. Even Bai et al. (2018) point to the beneficial role of deregulation since banks have shifted credit toward more productive young and small firms. However, recent work from Karakaya et al. (2022) suggests that these beneficial effects of deregulation may have been affected by the prior experience of banks serving specific sectors in their home state. However, even in this case, results are more robust for sectors that rely more on external finance, have lower amounts of physical capital that can be pledged as collateral, generate more valuable patents, and have a higher risk.

On the other hand, Kerr and Nanda (2008) demonstrate that the number of new firms increased following deregulation, but the churning rates of new ventures were also high. Most of the start-ups failed within three years of founding. Rogers (2012) investigates the impact of bank market structure differences within U.S. states on firm creation. States with smaller banks and those with more branch locations (but fewer banks) where presumably bank market power is high contribute more to new firm creation.

More recently, research has linked banking competition with firm innovation. Benfratello et al. (2008) show that local banking development in Italy plays a positive role in innovation. Kendall (2012) looks at the regional growth in India and finds that more bank credit is beneficial for local economies, mainly due to an improvement in human capital which is approximated by the local literacy rate of the population. Hsu et al. (2014), using a sample of 32 developed and emerging countries, find that industries that are more dependent on external finance and that are more high-tech intensive exhibit a higher innovation level in countries with better-developed equity markets, but the development of credit markets appears to discourage innovation in industries with these characteristics. Cornaggia et al. (2015) document a positive impact of banking deregulation in the U.S. on innovation for firms more dependent on external finance.

4.1.4 Banking Crises and Bank Bailouts

Banking crises and bank bailouts represent another type of natural experiment where researchers can observe the potential effects of a bank failure or bailout on the real economy. From a theoretical point of view, the seminal paper of Gertler and Kiyotaki (2010) was the first to introduce in previous macroeconomic models (Bernanke et al. 1999), the role played by financial intermediaries during a financial crisis in amplifying the effects of adverse shocks and in delaying recovery from them. Then the role of various possible central bank interventions aimed at mitigating the shock that distressed intermediaries face is examined. These actions should support banks in restoring lending to the real economy.

Several studies build on the seminal paper of Gertler and Kiyotaki (2010) by adding various levels of complexity and realism to their initial intuition. In Gertler and Karadi (2015), the central bank can use its unique ability to elastically raise funds to effectively replace at least part of the lost intermediation activity, lowering credit spreads and raising investment. Stein (2012) demonstrates that bank monitoring by a regulatory authority is necessary since the banking sector tends to produce more short-term debt than is socially optimal, which increases bank riskiness. Therefore, an ex-ante monitoring policy is necessary to restrain banks from engaging in this sub-optimal choice rather than intervening afterward through asset purchases.

In Brunnermeier and Sannikov (2014), firms experience large fundamental shocks that destroy their net worth, putting the economy in a crisis regime in which the adverse effects of the initial shock are persistent and in which small additional shocks can have considerable effects. He and Krishnamurthy (2013) demonstrate that funding constraints of intermediaries can cause the economy to enter into a crisis, because they hold risky assets in their portfolio, and a funding shock forces banks to increase leverage to the expense of equity. When the equity constraint starts to bind, the economy enters into a crisis region in which the relationship between real shocks and risk premia is much sharper and in which specific government policies employed during the crisis (such as equity injections into intermediaries and direct asset purchases) can produce stabilizing effects. Gertler and Kiyotaki (2015) integrate into these models bank runs. If banks issue only short-term deposits, when households ask for their redemption, they need to sell some of their risky assets. This exposes the banks to capital losses because of fire sale dynamics. If a banks' net worth is sufficiently low, it would be forced to liquidate all its assets exposing its depositors to potential losses. If the latter perceive that other depositors will bank-run, they will be incentivized to run as well, causing the financial system's collapse.

The regulators' reaction to bank distress has traditionally been to offer them support with public resources, given that the risk of bank runs and consecutively, a collapse of the financial system is very high. The empirical evidence on bank bailouts and growth has yielded inconclusive results. There are several reasons for that. On the one side, Claessens et al. (2005) or Dell'Ariccia et al. (2008) claim that bank bailouts do benefit economic output. Rather they could even generate moral hazard issues that hamper growth (Dam and Kötter 2012).

Laeven and Valencia (2013) and Dinger et al. (2022) extend the concept of bailout, by including the liquidity support provided by central banks to avoid liquidity pressures on banks' which could trigger insolvency risks on some of them. The former find that fiscal support positively impacts economic recovery. Conversely, liquidity support does not produce any tangible impact on growth.

Dinger et al. (2022) extend the work of Laeven and Valencia (2013) by looking at the sectoral value added in different countries, considering that fiscal and liquidity support are interrelated, and by investigating whether bank lending is the primary driver through which bank distress and bank bailouts drive economic growth. Their results emphasize that bank bailouts contribute to economic stabilization after a banking crisis. Moreover, they also indicate that monetary support has to be complemented by recapitalizations of the banking system to limit the risk that bailouts increase moral hazard and lead to the misallocation of credit toward less efficient firms.

4.1.5   Potential Nonlinearities of the Finance-Growth Nexus

Recent papers that delve more into the link between finance and growth have dedicated particular attention to developing nations by conducting a series of single-country analyses. Ang and McKibbin (2007) for Malaysia, Bandiera et al. (2000) for Ghana and Turkey, Arestis et al. (2002) for India find that financial liberalization measures as removals of capital controls, interest rate ceilings, direct credit programs, and high reserve requirements may improve financial development but also induce fragility in the financial system with an ambiguous impact on long term growth. This could explain the financial liberalization policies' failure in many developing countries in the 1970s.

Similarly, other authors have challenged the view that more finance means higher growth rates. The Global Financial Crisis demonstrated that malfunctioning financial systems could, directly and indirectly, waste resources, discourage saving, and encourage speculation, resulting in underinvestment and misallocating scarce resources (Law and Singh 2014). The impact of financial development on growth may not be linear. There could be a turning point in the effect of financial development on growth. This could be because the financial sector competes with the rest of the economy for scarce resources. Several papers have documented this nonlinear relationship (Cecchetti and Kharroubi 2013; Shen and Lee 2006; Huang and Lin 2009; Law and Singh 2014). Henderson et al. (2013) suggest that while the positive impact of financial development on growth has increased over time, it is also highly nonlinear, with more developed nations benefiting while low-income countries do not benefit at all.

Beck et al. (2014) corroborate these findings by evidencing that when the financial sector increases in size but departs from intermediation activities and engages in other services, its effect on the economy is very weak and increases overall growth volatility. In advanced countries, an expansion of non-intermediation services and the corresponding increase in the overall financial sector's size impacts positively on growth but comes at the cost of higher volatility.

Figure 4.1 summarizes the most important variables at play in this vast literature. The financial sector's activity is proxied through different variables which look at credit and stock markets, as mentioned at the beginning of the section. Several studies have considered the financial sector's size a moderator rather than a proxy of financial development. Growth initially has been investigated at the country level, and then with more data availability, researchers have adopted an industry or firm-level lens.

The moderating variables at the country level are related to each country's current legal and institutional framework. Bank ownership has attracted significant interest together with bank deregulation (in the U.S. mainly) and bank market power. These are jointly important with firm characteristics in each country/region (dependence on external finance, degree of opaqueness, innovative efforts).

4.2   Do Banks Lend More
and to Whom After a Natural Disaster?

This section looks at the theoretical and empirical evidence on bank lending behavior after severe natural disasters. These contributions build on the previously described streams of research since disasters are considered exogenous shocks to bank activity. It is also worth noting that bank lending is affected by other internal or external factors to the banking institution. After revising the mechanisms through which weather

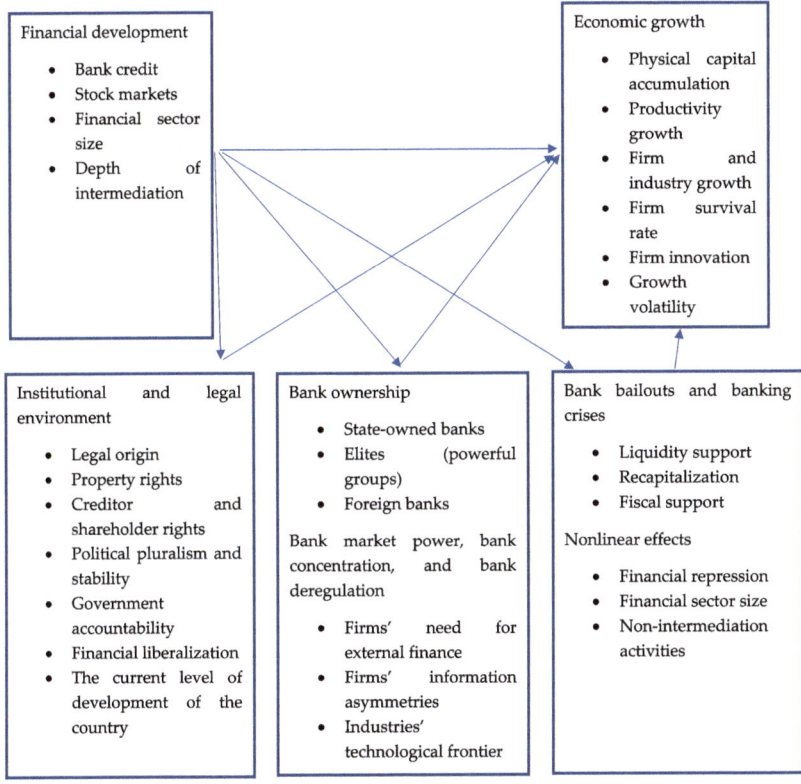

Fig. 4.1 Empirical findings of the finance-growth literature and main variables at play

events impact bank lending, the focus will shift toward banks' impact on post-disaster economic growth.

Natural disasters, as mentioned previously, cause a local shock to areas stricken by them which could impact the financial sector's operations, asset liability management and risk. At the same time, demand for credit will increase in order to rebuild damaged properties and restore production and regular activities for businesses. Disasters could also propagate to nearby or distant areas through spillover effects.

Different moderating factors are important in this context. These comprise lender type (i.e., local vs. nonlocal), lending technology (relationship vs. arm's length lending), countries' level of development (advanced vs. emerging), bank capital adequacy, presence of securitization markets, or bank market structure. Except for a few cases, this literature is relatively recent but is growing, indicating the researchers' interest in bank activity after severe weather hazards.

The first study linking natural disasters with the financial sector is Odell and Weidenmeier (2004), who look back to the earthquake that hit San Francisco in 1906. It generated massive losses amounting to around 1.8% of U.S. GNP. Claims paid by British insurers to those who suffered losses from fire outbreaks after the earthquake were enormous. They generated an outflow of gold in the early autumn 1906 and forced the Bank of England (followed by central banks in France and Germany) to stop discounting American finance bills for the following year. This generated a sharp but short-term recession in the U.S. during 1907.

Berg and Schrader (2012) is the first theoretical paper linking a major natural disaster to borrowers' access to bank credit. They analyze the effect of an unpredictable aggregate shock deriving from the eruption of a volcano in Ecuador to loan demand and access to credit post-disaster. The theoretical model introduces essential elements that characterize borrower–lender relationships following a catastrophic natural event observed even in more recent studies. The authors benefit from a proprietary dataset of loan-level data obtained by a large microfinance institution in the country. Microfinance institutions establish long-lasting relationships with borrowers in emerging economies to overcome problems of asymmetric information and lack of collateral. In the banking literature, this is commonly considered relationship lending and has been thoroughly studied both from a theoretical and an empirical point of view.[2]

In the aftermath of a natural disaster, the shock can be reflected in the reduction of entrepreneurs' physical capital (direct damage) and loss of revenues in the short and long term through macroeconomic shocks affecting the firm (feedback loops). These damages are assumed not to

---

[2] Relationship lending can be defined as a long-term implicit contract between a bank and its client (Boot 2000). Relationship lending can be distinguished from asset-based or transaction lending, where the lending decision of a bank depends respectively on the value of the clients' collateral, or credit scoring (Von Pischke 2002).

affect the liquidity of the bank and its overall stability. The losses incurred by the entrepreneurs will push the latter to increase the required credit. The bank has to tighten lending standards to obtain a client pool with higher average credit quality. Consequently, as the credit quality of the loan applicants remains unaffected by the shock, the need for higher lending standards will result in a refusal of loan applications. Especially low-quality firms and those strongly affected by the shock will lose access to finance. Therefore, a decrease in the fraction of credit applicants receiving a loan is observed. However, the bank can screen well from bad borrowers by looking at their repayment rate and, therefore, will not restrict access to repeat borrowers (relationship borrowers).

The empirical results confirm that demand for credit increases significantly after volcanic eruptions, suggesting a need for post-disaster additional financing exists. The results indicate that high volcanic activity in the last months before the credit application led to significant decreases in the probability of loan approval. However, returning clients have a higher probability of receiving a loan in general and are about equally likely to be approved for a loan after volcanic shocks occur. The bank faces a high degree of asymmetric information and cannot adequately screen good from bad entrepreneurs, therefore is forced to refuse credit to a large number of (new) applicants.

4.2.1 *Differences in Post-disaster Lending Behavior Between Local and Nonlocal Banks in the U.S.*

The most considerable bulk of research on post-disaster lending is related to different patterns of lending strategies adopted by banks that concentrate lending in areas affected by the disaster and other institutions which operate nationwide and are more diversified. This literature deals mainly with the U.S. context since in this country, after the banking deregulation waves described previously, two types of banks co-exist, namely multimarket banks that operate in more than one state through an extensive branch network and concentrated local lenders whose activities are limited inside a single area (local market banks).

This distinction has posed to researchers the dilemma of which type of bank is more entitled to lend after a severe shock such as a natural disaster. On the one side, local banks should be more vulnerable to losses of income arising from the shock. Therefore, they should tighten credit standards for borrowing firms and households. Hence, lending should originate from large diversified lenders (*financial capacity channel*) (Chavaz 2016).

On the other hand, by adopting relationship lending technologies, local banks could have an advantage over multimarket institutions, since the former can reduce information asymmetries and better screen, monitor and price new loans despite *depressed* or uncertain collateral values. They also have the incentive to lend more if this contributes to increasing local house prices in an area where the bank has a significant market share or that represents a large portion of the loans held in the portfolio (*relative profitability channel*) (Favara and Giannetti 2015; Chavaz 2016).

The U.S.-focused studies generally look at the impact of hurricanes or tropical storms since they are largely unexpected, exogenous to the socio-economic environment and cause extremely high damages and losses to households and businesses. Chavaz (2016) studies bank lending during the protracted recovery from the 2005 hurricane season, the costliest natural disaster in recorded U.S. history. Hurricanes Katrina, Rita, Wilma, and Dennis damaged 1.19 million housing units, a large part of which were insufficiently insured. The econometrical setup allows us to identify the causal impact of a downturn on credit allocation for differently diversified banks and thus disentangle the two conflicting channels of interest. The results confirm that the relative profitability channel is predominantly at work. A 10% higher share of bank branches in affected counties is associated with a 9.2% higher (log) lending growth in affected counties. Local banks are most prone to accept information-intensive mortgage applications in affected counties, confirming the theoretical view that they have a comparative advantage over other diversified lenders due to the relationship lending approach they adopt in lending.

Gallagher and Hartley (2017) investigate this problem from the point of view of households that Hurricane Katrina hit New Orleans. They find that overall household debt was reduced without a remarkable impact on loan delinquencies or credit card borrowings. Borrowers used flood insurance or government aid to repay existing loans rather than to rebuild. Mortgage reductions were more significant in areas where reconstruction costs exceeded pre-Katrina home values, and mortgages were likely to be originated by nonlocal lenders. Here the incentive problem mentioned by Chavaz (2016) is again observable. First, the success of banks with an extensive lending presence in New Orleans is highly dependent on the continued borrowing of New Orleans residents and on the overall economic well-being of the city. Lenders that have a relatively small share of their business in New Orleans may prefer to protect themselves from the uncertain economic environment of post-Katrina New Orleans by reducing their lending exposure and allocating capital elsewhere. Secondly, since local lenders were likely to have more personnel based in New Orleans, the costs to monitor the reconstruction process and protect their investment would be more mitigated than nonlocal lenders. Consistently with these expectations, Gallagher and Hartley (2017) find that nonlocal lenders dramatically reduced their lending to New Orleans (relative to local lenders) after Katrina.

However, when reconstruction activities increase credit demand, large banks could be better positioned to taping this opportunity given their possibility of getting funded through a variety of sources which local banks usually do not possess. Large banks benefit from an internal capital market, through which they can move funds toward the most deserving projects and, due to external financial constraints, away from less deserving ones (Stein 1997).

Cortés and Strahan (2017) look at the response of large banking institutions to the increased demand for credit after severe hurricanes in the U.S. from 2001 to 2010. They find that these banks eventually compensate for this increased non-expected credit shock by reducing their exposure to other regions across the country. The reduction is observed only for those counties where the bank does not operate through a branch. Therefore, the decline in lending is concentrated outside banks' core markets, where the bank cannot produce valuable services such as access to superior information from borrowers or monitoring insofar as their credit could not be provided on equal terms by other banks or intermediaries. The findings are coherent with other prior studies suggesting that a bank's physical presence in a market improves access to information about borrower quality and the value of collateral, allowing the bank to earn rents (Gilje et al. 2016). The increased lending from large banks does not contradict the relationship lending view since banks do not cut lending in areas where they are located nearby their customers.[3] Moreover, banks exposed to natural disasters increase rates on deposits across

---

[3] Local lenders extend more credit to opaque (small) borrowers than distant ones. Loan rates tend to decline with borrower–lender distance. In mortgage finance, locally concentrated lenders focus on soft information-intensive segments of the mortgage market and have an advantage in screening and monitoring riskier borrowers (Gilje et al. 2016).

other markets where they own branches to help fund the unexpectedly high loan demand in the shocked markets.

The paper of Cortés and Strahan (2017) documents that diversified banks increase credit in affected areas by hurricanes by retracting their activities in other non-core markets. This finding is interesting since it shows that disasters could potentially have spillover effects in other areas of the country via the bank network channel. A similar outcome is observed by Ivanov et al. (2022), who find that credit in non-affected areas decreases post-hurricane strike. This is observed mainly for low-capital banks. A similar result contrasts with the benefits of interstate banking documented by Jayaratne and Strahan (1996) and other studies. However, the U.S. banking market partially compensates for the credit reduction through shadow banks which serve primarily the term loan syndicate market but do not tap into the credit lines one.

The drop in credit line facilities could be detrimental to firm activity since they are a vital source of cash for firms when they face adverse shocks. Brown et al. (2021) find that profitable small firms use bank credit lines extensively in times of unexpected shocks such as abnormal heavy winter snowfall. Credit lines are used as liquidity insurance against cash flow volatility.

Duanmu et al. (2022) investigate how bank residential mortgage lending standards are affected by risks to the local economy from natural disasters. Banks tighten lending standards in disaster-hit counties, suggesting that lenders are more cautious in these locations since environmental disasters can increase the long-term risks to the local economy and the bank's stability. Their outcomes are driven by large and diversified banks, which have tightened credit standards. In contrast, local lenders can still identify creditworthy applicants using soft information and continue to lend after natural disasters. The findings are also consistent with other recent work showing how banks are starting to price differently and charge higher interest rates for long-term loans to firms in areas prone to physical risks (Jiang et al. 2019).

The mortgage market in the U.S. is becoming more variegated since nonbank lenders (shadow banks or FinTech lenders) are increasingly competing with traditional banks for business. FinTech lenders have emerged as real disruptors of the mortgage application and underwriting processes. Allen et al. (2022) investigate whether technology-equipped lenders can fill the credit gap when a local mortgage market faces unanticipated and temporary demand pressure from the reconstruction of damaged property following natural disaster shocks. Unlike traditional banks, FinTechs adopt fully automated algorithms that integrate machine learning and artificial intelligence to process especially hard credit information more efficiently than traditional lenders.

Several interesting findings emerge from this study. First, traditional and FinTech lenders increase credit supply following a weather shock. This does not come at the expense of charging higher rates to borrowers from FinTech lenders. There is no higher likelihood of delinquency on post-disaster fintech loans. The post-disaster shift in credit supply for FinTech lenders emanates from counties where traditional banks depend on on-balance sheet lending and physical branch networks. FinTech lenders have a competitive advantage in these markets due to their reliance on securitization and online access. Moreover, FinTechs respond to a highly competitive market by aggressively relaxing underwriting standards.

This study shows that the banking landscape in developed economies like the U.S. is transforming and relying heavily on technology and artificial intelligence. This will considerably affect the market structure, the industrial organization of banks and other operators, and how they respond to exogenous shocks such as catastrophic weather events.

4.2.2   Empirical Evidence from Cross-Country Studies and Other Economies

Outside the U.S., several contributions look at the role of banks in providing recovery lending after major natural disasters and how bank performance is eventually affected by these catastrophes. Collier and Babich (2017) perform a cross-country analysis for a period of 18 years over 78 countries. They first develop a model to identify the drivers of bank lending reduction after a disaster, which could be either capital scarcity (financial institutions may adjust lending because disasters erode its capital through an increase of loan losses) or reduced return (disasters can affect the productivity of bank's borrowers and increase their probability of default thus increasing credit risk for the bank). The results show that loan growth decreases after severe disasters. Disaster-related contraction is due to a reduction in lending from financial institutions with low pre-event capital ratios. Both low- and high-capital banks lend less following disasters in economies with low insurance coverage, suggesting that reductions in the expected returns on lending help explain these disaster-related credit declines.

Thuy Nguyen et al. (2023) empirically look at the impact of natural disasters on commercial bank performance and how financial integration moderates this relationship for a sample of East Asian countries over the period 1999–2014. Their findings confirm that local banks suffer a reduction in their deposit ratios since customers redeem their claims due to financial constraints. However, disasters do not contemporaneously affect bank liquidity, credit risk, profitability, and default risk. Furthermore, foreign banking claims, specifically those extended by regional Asian lenders, help to alleviate the deposits' decline in the aftermath of natural disasters. This corroborates the thesis that financial integration, especially intra-regional, is beneficial for banks since institutions located in neighboring countries have an informational advantage over other foreign banks domiciled farther away.

Several single-country studies look at commercial banks or microfinance institutions (MFIs), which provide credit to the poorest cohorts of populations in emerging markets. The latter generally evidence the positive role of MFIs in alleviating credit constraints after major natural disasters and stresses the importance of funding MFIs by foreign donors in these situations to allow them to refinance borrowers seriously damaged by the calamity (Kianersi et al. 2021; Czura and Klooner 2023).

Regarding the role of commercial banks, Celil et al. (2022) examine the role of City Commercial Banks (CCBs) in alleviating financial constraints for those affected by natural disasters in China. These are state-owned institutions that concentrate lending in a restricted geographic area and also have social goals due to state ownership. CCBs expand credit aggressively in cities that experience natural disasters with a substantial adverse economic impact. Moreover, CCBs with more significant post-disaster credit expansion experience a decrease in non-performing assets in the subsequent years, rejecting the view that the increased lending coincides with loosening credit standards. Therefore, concentrated lenders similar to what was observed in the U.S., can respond positively to natural disasters by increasing lending without compromising their stability.

Nguyen and Wilson (2020) assess the impact of the Indian Ocean Tsunami of 2004 on the aggregate credit supply to Thailand provinces. The results of their investigation suggest that the tsunami had long-lasting adverse effects on bank lending, albeit the effects were spread unevenly across geographic areas, with most of the reduction in aggregate lending occurring in severely affected provinces. The presence of bank branches in affected regions mitigates some of the adverse lending effects that follow the tsunami. Even in this case, bank branching is beneficial since it helps to overcome asymmetric information issues of creditworthy borrowers.

Another empirical study looks at the German case in the aftermath of catastrophic floods that hit the southeastern part of the country in 2013. Koetter et al. (2020) show that local banks provide corporate recovery lending to firms affected by adverse regional macro shocks due to damage and losses from flooding. At the same time, there is little evidence that solvency or credit risk among affected banks worsens relative to their unaffected competitors. However, regional banks benefit from being part of geographically diversified conglomerates.

4.2.3   *The Link Between Post-disaster Recovery Lending and Economic Recovery*

The empirical evidence in the previous section generally acknowledges that banks increase lending after natural disasters, conditional on several factors. However, studies documenting how this lending might translate into potentially higher economic growth are less numerous. In this section, we aim to present the main findings of this bulk of studies and the main drivers that foster economic recovery through the bank lending channel.

From an econometrical point of view, natural disasters can be considered natural experiments that could generate simultaneous shocks in the demand and supply of credit. Therefore, researchers are faced with the challenging task of estimating two separate channels: *the bank lending channel*, i.e., the inability of banks to cushion borrowing firms against banks' exposure to losses from natural disasters, which reduce bank credit supply, and the *firm borrowing channel*, i.e., the borrowers' performance (which is impacted from the disaster) influences the bank's decision to extend credit in the first place.

Other past contributions have circumvented the difficulty in simultaneously estimating the bank lending and firm borrowing channels differently. Peek and Rosengren (2000) examine whether construction activity in the U.S. was affected by the deterioration of Japanese banks during the 1990s through a reduction in lending of these banks at their

U.S. branches. Similarly, Khwaja and Mian (2008) examine the transmission of a bank liquidity shock to client firms in the wake of unexpected nuclear tests in Pakistan.

Hosono et al. (2016) use a natural disaster as a shock to the loan supply but not to the loan demand. Their goal is to examine whether damage to banks adversely impacted the investment of client firms that did not themselves suffer any damage. The focus is on the behavior of banks and borrowers located in Japan's region affected by the Kobe earthquake of 1995. Banks could cut lending and affect borrowers' performance after a natural disaster for two reasons. On the one side, banks could face severe damages and losses, which could compromise their ability to correctly screen borrowers' creditworthiness, reducing their ability to originate loans. On the other, natural disasters could damage borrowing firms, which would make it difficult to repay loans and lead to a deterioration in the banks' loan portfolio.

They also examine how damaged firms recovered after the earthquake, i.e., whether damaged firms increased capital investment relative to undamaged firms and whether damage to banks negatively affected damaged firms' investment. The findings demonstrate that unaffected firms outside the earthquake area, associated with an affected bank, reduce investment after the disaster. The exogenous damage to banks' lending capacity had a significant adverse effect on firm investment.

Rebhein and Ongena (2022) adopt a similar econometric strategy as Hosono et al. (2016) by focusing on the link between recovery lending and growth after the floods that hit Germany in 2013. They first identify firms in non-flooded areas that are associated with disaster-exposed banks and compare them to firms located in the same region but not connected to a disaster-exposed bank. On average, banks' lending shifts from non-disaster regions to disaster regions reduces borrowing by 2.4%, employment by 2.4%, and tangible assets by 5.1% for firms connected to a strongly exposed bank. This shift is driven by undercapitalized banks. Another interesting finding is that firms in regions with higher ex-ante disaster risk suffer disproportionately large real effects. It suggests that banks rebalance their portfolio by reducing their exposure to areas with a high chance of floods in the future.

In an already mentioned study, Celil et al. (2022) look at the role of regional lenders in promoting economic recovery in China. They find that concentrated lenders (City Commercial Banks) contribute significantly to the revival of economic activity through the lending channel in the cities they operate. This post-disaster growth is measured through national GDP statistics but also "captured" by the city night lights. This latter measure can be considered a clean measure of the city's growth compared to official statistics which could be inflated by government officials who have political incentives to post aggressive growth numbers following disasters.

In other works focused on local vs multimarket banks in the U.S., there is evidence that their recovery lending activity helps job retention and creation at young and small firms. Cortés (2014) finds that the direct effect of natural disasters on firm growth is negative, but the interaction with local banks' market share is positive and significant for new firms, firms 2–3 years old, and firms that are more than 11 years old. The growth rates of young and mature firms' employment share increase respectively by 1% and 0.1%, when local credit is one standard deviation higher than the mean. Similar results are observed for employment growth by firm size. This is a consequence of recovery lending by local banks. Cortés (2014) finds that these institutions maintain the increase in lending for up to two years after the disaster.

Schüwer et al. (2019) explore bank behavior in the areas hit by Hurricane Katrina in 2005. Initially, they delve at the link between disaster and bank capital ratios and find that only the best-capitalized banks that do not belong to a banking group increase their capital ratios after this catastrophic event. The rationale is that these banks bear high bankruptcy costs and a high franchise value which they want to conserve compared to pre-hurricane less capitalized banks. Highly capitalized independent banks also increase their new lending to non-financial firms in their core markets (where they have a branch presence). Therefore, these banks reduce their loan exposures by not reducing new lending but through other strategies such as loan sales.

In addition, Schüwer et al. (2019) test whether the presence of independent, well-capitalized banks fosters economic recovery in the counties where they are domiciled. The empirical outcomes confirm that counties with a higher share of independent banks and relatively high average bank capital ratios are associated with better economic growth in total personal income and employment than other counties (with fewer independent banks or lower average bank capital ratios). This finding is consistent with the evidence produced by Rebhein and Ongena (2022) about the importance of bank capital buffers for fostering economic recovery after a significant weather shock.

In another recent paper, Duqi et al. (2021) focus on the real effects of lending at a sample of U.S. counties stricken by major hurricanes in the period 1995–2017. Their study contains various innovations compared to the previous research. First, they consider the importance of bank market structure since previous research described above indicates that banking competition is strongly related to bank lending, and there could be two alternative economic channels at work (see Sect. 4.1). Second, they focus explicitly on county rather than firm-level growth, as other papers have attempted to do previously. Third they investigate whether economic growth after a natural disaster results from commercial rather than residential mortgage lending.

The micro-level evidence indicates U.S. counties with less competitive banking sectors display higher post-disaster growth rates. Following a hurricane, a 10% reduction in banking competition leads to a statistically significant 0.32% increase in the rate of economic growth. The effect of market power on the recovery is most pronounced in those counties where profitable and better-capitalized banks operate, suggesting that these banks deploy the retained earnings they accumulate during normal times to support the local economy after a disaster.

Second, they find that market power allows financial institutions to support distressed borrowers to prevent foreclosure by renegotiating mortgage terms and refinancing existing mortgages. There is no link between market structure and consumer or commercial and industrial (C&I) lending after a disaster. In these segments, banks intermediate disaster loans on behalf of the Small Business Administration (SBA). The supply of SBA disaster loans is invariant to the bank market structure because it is the SBA that evaluates and approves loan applications. The importance of market power for banks relates to their ability to extract rents from their borrowers in "good times."

The beneficial impact of a low competitive banking market on local economic recovery is associated with a larger share of small banks in those counties which engage in traditional banking activities. Through their existing borrower relationships, small lenders have superior information about a customer's actual credit risk, which allows them to quickly evaluate the value of destroyed collateral and extend credit to affected households by quickly assessing their repayment capacity. Thus they have an advantage in managing riskier loans due to their screening and monitoring abilities. This outcome is observed in other studies mentioned previously, such as Petersen and Rajan (1995), Loutskina and Strahan (2011). It highlights the positive side of relationship lending (and bank market power) after a severe weather shock.

The U.S. financial market is highly integrated and has benefitted significantly from the advent of securitization since the 1970s when two Government Sponsored Enterprises (GSEs), namely Fanny Mae and Freddie Mac, were created. Their core activity is to guarantee and purchase mortgage loans from commercial banks. GSEs' access to implicit (now explicit) government support allows them to borrow at rates below those available to private banks and to offer credit guarantees on better terms than competitors without such implicit support (Loutskina and Strahan 2015). GSEs purchase only mortgage loans that fall below a specific limit, which is exogenously based since they aim to promote access to mortgage credit for low and moderate-income households.

In times of unexpected shock to credit demand, such as in the aftermath of severe natural disasters, numerous papers have documented the positive role of securitization markets in the U.S. Local banks that appear the most active in granting credit access to households after the disaster tend to securitize a large part of their mortgages. This occurs because they face financial constraints due to loss of income and damage from the catastrophic event, which limits their capacity to lend. Selling loans in the secondary markets allows them to transfer the risk to diversified agents and, at the same time, exploit their competitive advantage in lending to affected markets (Chavaz 2016).[4] Cortés and Strahan (2017) reach similar conclusions and notice that banks use securitization/sales to substitute for on-balance sheet finance required to lend in shocked markets, thus mitigating (partially) the need to cut loan originations in connected markets.

Since securitization increases the propensity of lenders to transfer disaster risk to GSEs, this could generate opportunistic behavior in banks and weaken the discipline brought about by the mortgage finance industry in fostering climate change adaptation. Ouazad and Kahn (2022) find that banks tend to originate and sell loans just below the threshold, which qualifies non-jumbo (i.e., that can be sold to GSEs) from non-jumbo ones after extreme disasters. After a billion-dollar catastrophic

---

[4] The GSEs enhance liquidity by buying mortgages directly from lenders and also by selling credit protection that allows such mortgages to be securitized easily by the originator (Cortés and Strahan 2017).

event, the probability of securitization increases by up to 19.3 percentage points. The evidence points to an increase in adverse selection after natural disasters. Securitizers could respond by pricing climate risk in their guarantees, and integrating their disaster risk models with disaster risk data.

Figure 4.2 summarizes the main empirical findings of the literature on the impact of natural disasters on bank lending and growth. A few key elements emerge from it. First, local banks are more active in addressing the need for credit from businesses and households arising after a natural disaster. They are interested in "protecting" these borrowers in hard times since their lending is concentrated in a specific region. They also possess the skills and personnel to evaluate local borrowers' default probability and damaged collateral's value correctly. At the same time, they have a high degree of market power, allowing them to extract rents in "good" times. However, this is not enough. These banks need to be adequately capitalized. They also benefit from a well-functioning securitization market in the U.S. Outside the U.S., local banks' positive role is supported by being part of large conglomerates (as in the German case) or by international funding from regional lenders as in Asia. Last, the role of government is essential, since state-owned agencies provide subsidized credit and grants. In the U.S., subsidized loans are mainly addressed to business borrowers. Therefore banks foster recovery through residential lending primarily. This market is now witnessing the rise of alternative players, such as FinTech banks, which are aggressively gaining market share at the expense of traditional banks. FinTech banks benefit from a highly efficient business model and the securitization of their portfolio.

4.3   How Can Banks Incorporate Climate Risks in Their Lending Strategies?

The empirical outcomes evidenced in the previous sections indicate how bank lending patterns change after extreme weather events. This is observed mainly for concentrated banks. However, even diversified ones and other new players may choose to respond to the increased demand for credit, allocating funds away from non-core markets. Conversely, acute physical risks (such as natural disasters) or chronic ones (such as sea level rise or chronic heat waves) will increasingly impact bank risk through different trajectories. Therefore, banks' lending strategies can no longer avoid the impact of increased climate risk on bank performance.

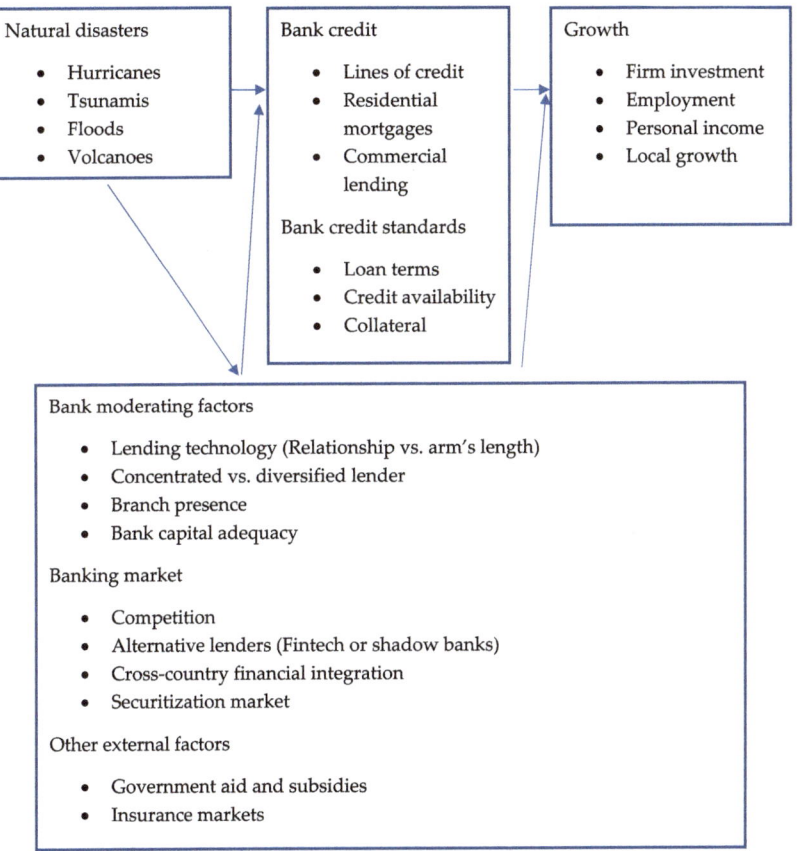

Fig. 4.2 Main empirical findings and variables at play in the literature on the impact of disasters on bank lending and growth

That is why climate risks should be integrated into banks' strategic planning. The ECB has required banks to factor in the impact of climate risk in the short, medium, and long-term business strategy to take the necessary corrective actions.

The current business models should be stress tested over the long term to assess their resilience to various alternative climate risks under different climate scenarios. Banks are encouraged to produce a series of KPIs for each climate risk they deem relevant, which should be measurable.

Figure 4.3 highlights the areas that should be involved in finalizing commercial banks' climate risk-oriented strategic and operating process. Although the focus is on catastrophic weather events classified as physical risks by financial intermediaries and their regulators, the framework described in Fig. 4.3 should also consider transition risks.

From a strategic planning perspective, bank customers are grouped under different criteria based on the banks' business model, such as by setting revenue thresholds, industrial sectors, etc. It is not agile for banks to categorize these groups based on their exposure to certain climate risks. Therefore, a more detailed breakdown of specific industries would be necessary. Banks should also adjust their internal control and management accounting systems to correctly identify borrowers' exposures to these risks.[5]

Fig. 4.3 Banks' strategic areas which will be impacted by climate risks

5 AIFIRM Position Paper 39/2022.

Consequently, new pricing models should be adequately introduced to reflect the different riskiness emerging from this new categorization. Reports to different internal and external stakeholders should be updated based on reliable data that should frequently incorporate new information about customers' climate risk exposure.

At this stage, investors are assigning increasing importance to specific sectors' exposure to certain climate risks. All industries related to infrastructure development in disaster-prone geographic areas, as well as real estate and commodities, will be increasingly screened based on their exposure to physical risks. In portfolio asset allocation, asset managers are introducing factors related to climate risks that could modify their investment choices. However, these asset allocation techniques are based on historical time series, which could be insufficient to predict future losses under different climate change scenarios.

The asset management industry and financial institutions with a large amount of investment-grade securities in their portfolios face two types of difficulties in incorporating climate risks in their asset allocation strategies. First, there are few "extreme events" since observability of catastrophic natural disasters is rare. This generates an estimation error. Additionally, the dynamic process through which climate risk affects other assets and how they spread in the long term yields an even more problematic estimation of the return from these assets (Bolton et al. 2020a). For example, the thaw of permafrost in Siberia could make this region more populated and more fertile, but at the same time, could generate extreme droughts and hunger in southern Europe.

Regarding credit origination, banks could follow two types of strategies. Under the so-called *top-down* approach, banks could concentrate lending toward those sectors/geographies/companies which are less impacted by certain risks, or that can be considered really "green" and deleverage gradually versus the others. The *bottom-up* approach instead is focused on the single exposure/customer. The analysis should rely on a sound credit risk management of physical risks and needs to take into account the fact that this type of risk impacts firm assets through damages to their production facilities, but also indirectly through an interruption of the firm's supply chain if along this chain certain suppliers are exposed to the same risk. Bank should also evaluate the exposure of real estate guarantees to climate risks. This analysis needs a robust model that can evaluate physical risk exposure in real-time and the relative risk-adjusted return that the bank should earn through correct product pricing. Banks should start collecting data specific to each customer through detailed surveys during the credit origination process about the resilience of customers to climate risks.[6]

Following credit origination, monitoring banks' exposure to certain climate risks is necessary. This could change the bank's strategies toward specific sectors or geographic areas and allow bank departments to take necessary actions to avoid negative implications on bank profitability/ stability stemming from unexpected climate risks (EBA 2020). These implications could include a sharp drop in productivity or profitability of specific sectors/firms, an abnormal increase of non-performing loans following climate events, or the introduction of new technologies that can mitigate climate risks in specific geographic areas. Indications from these monitoring actions should foster changes in annual budgets and industrial plans if these climate alternations are deemed permanent.

Successful monitoring needs reliable KPIs related to specific climate factors.[7] The World Economic Forum has generated a set of KPIs since 2020 which should capture companies' efforts toward sustainability and non-financial disclosure. These indicators should help stakeholders compare different companies regardless of their industry or region. The metrics include non-financial disclosure data centered around four pillars: people, planet, prosperity, and governance principles. Indicators related to climate change include air pollution, nature loss, and greenhouse gas emissions, among others.[8]

The European Banking Authority has reminded banks that when considering introducing a dedicated treatment, an understanding of the likelihood that environmental risk drivers might already be reflected in the prudential regime is needed. This ensures that environmental risk factors are appropriately captured, avoiding underestimation or double counting, which would weaken the consistency and robustness of the prudential framework. In other words, climate risks could be already indirectly incorporated into bank credit risk models through their impact on

---

[6] An example of customer resilience to damage from hail is an investment to strengthen the roof of firms' production facilities.

[7] Mortgage finance could be related to a specific KPI showing the number of households that improve their building energy standards after receiving a mortgage loan.

[8] The complete set of metrics can be accessed at this weblink (https://www.weforum.org/reports/measuring-stakeholder-capitalism-towards-common-metrics-and-consistent-reporting-of-sustainable-value-creation).

the macroeconomy. Banks, therefore, need to assess through their climate risk analysis, if the latter are correlated with traditional risk metrics such as the probability of default or loss given default. It would be helpful at this stage to have an accurate indicator of an asset that is considered "green risk-free" by the market. It could serve as a benchmark against which to price other assets which incorporate higher climate risks in their price.[9]

Since climate risks impact the most on credit risk exposure, banks should carefully monitor their sectoral and geographic loan concentration. The ECB has stressed that portfolio exposure to physical risk should especially evaluate concentration in areas prone to severe natural disasters such as floods, drought, or hydrological instability (ECB 2020). Credit risk models, therefore, should continuously incorporate and map geographic exposure to relevant physical risks and the relative exposure of the bank to these areas.

Reassuming, banks could adjust its strategy by adopting a multi-scenario framework. This means that the baseline scenario could be enriched by assuming alternative, more pessimistic, or realistic climate change scenarios in line with the most updated scientific opinion on climate change. Through an interactive process, different scenarios are assumed to influence different bank profitability and stability drivers under a given set of constraints, such as maximum and minimum exposure to certain classes of assets, industrial sectors, geographic areas, product, and customer types. Different paths of portfolio evolution could be built, and the bank's top management could opt for the most suitable one given its risk-return combination. However, this could be complex since it needs very advanced risk models that should incorporate different sets of indicators and variables at the firm, industry, and country levels. A few banks so far have embarked on a similar strategy to incorporate climate risks in their asset allocation strategy.

Regarding the regulators' response, different views have emerged about incorporating climate risk in banks' regulatory capital. There could be a penalty toward more exposed industries or geographic areas, and then regulators should monitor how credit toward firms and households belonging to the former will evolve. If risk weights toward these entities are going to increase, the weighted average cost of capital for the bank would be higher if banks lend to them so that they would curtail credit.

---

[9] An example of a "green risk-free" asset would be the interest rate of green bonds issued by large corporates with a triple A rating.

However, the empirical evidence so far has demonstrated that this could depend on other factors, such as banks' level of capitalization. Highly capitalized banks can extend credit at better terms since market financing alternatives are available for them (Bridges et al. 2014).

Besides, bank credit supply is influenced by other factors such as borrowers' history, guarantees offered, economies of scope extracted from the borrowers by offering them other services, etc. Banks could eventually pass through the higher financing costs to these borrowers, although this would depend on the overall volume of exposure toward the "climate risky" customers. However, this would not translate into less credit for them automatically if these firms, which could be large corporates, have alternative sources of financing and, therefore, could replace bank credit with other sources (Carrascosa 2021). As a consequence, the desirable effect (cutting GHGs, for example) would not be achieved.[10] Regarding more specifically, climate exposure to acute physical risks such as natural disasters, the increased risk weighting would certainly damage mostly SMEs located in certain regions, which might not have adequate financing alternatives compared to larger counterparts. Moreover, cutting credit to industries located in some geographic areas would harm banks by increasing the risk of default for these borrowers, which could be already highly leveraged with the banking sector.

The alternative could be pricing more climate-friendly companies at favorable rates. As previously mentioned, there is no guarantee that a lower cost of financing for banks would automatically translate to the borrower through better financing terms since other factors, such as the borrower's credit history, guarantees offered, etc., would matter. If the market is engulfed with greener start-ups, banks might consider these companies as riskier and thus refuse to grant credit to them.[11] Again, in

---

[10] This could be why several authors claim that imposing some kind of Pigouvian tax on certain pollutant sectors would be a better way of achieving the desired climate goals (see for example IMF Fiscal monitor 2019 at this weblink [https://www.imf.org/en/Publications/FM/Issues/2019/10/16/Fiscal-Monitor-October-2019-How-to-Mitigate-Climate-Change-47027]).

[11] Carrascosa (2021) reports that the default level of greener companies is higher in Spain compared to more pollutant ones, probably due to their smaller size and younger age.

the case of physical risks, it is not clear how this measure could be implemented since all households and businesses in a specific region would be indiscriminately affected.[12]

Regulators should also transmit the message to the banking sector that any measure toward a more sustainable economy would be permanent. Otherwise, banks would not be so receptive to modifying their lending strategies.

Central banks and policymakers are facing the dilemma of imposing a higher cost of credit to borrowers exposed to certain climate risks. The effects of this choice on banks and the real economy are ambiguous. Doing nothing exposes all economies and banks over the long term to a greater risk of exposure to losses from specific sectors and hazards. Imposing additional capital requirements related to climate risks does not guarantee that the final effect will be achieved, especially if it is related to reducing greenhouse gases, for example. At the same time, a careful evaluation of the impact of these choices on growth is needed since it is essential for the success of the policies. Other measures in place that may be as or more effective than changes in capital requirements, such as introducing green performance indicators or special taxes, are more efficient but come at a high political cost.

REFERENCES

Acemoglu, D., Johnson, S., Robinson, J., 2005. Institutions as the fundamental cause of long run economic growth. In: P. Aghion and S. Durlauf (eds.), Handbook of Economic Growth, pp. 385–472.

Acharya, V.V., Imbs, J., Sturgess, J., 2011. Finance and efficiency: Do bank branching regulations matter? Review of Finance, 15, pp. 135–172.

Allen, L., Shan, Y., Shen, Y., 2022. Do FinTech mortgage lenders fill the credit gap? Evidence from natural disasters. Journal of Financial and Quantitative Analysis, pp. 1–42. https://doi.org/10.1017/S002210902200120X.

Andrianova, S., Demetriades, P., Shortland, A., 2012. Government ownership of banks, institutions and economic growth. Economica, 79, pp. 449–469.

Ang, J.B., McKibbin, W.J., 2007. Financial liberalization, financial sector development and growth: Evidence from Malaysia. Journal of Development Economics, 84, pp. 215–233.

---

[12] Banks could impose to borrowers additional capex to become more resilient to certain natural disasters such as floods, hurricanes, wildfires etc.

Arestis, P., Demetriades, P.O., Fattouh, B., Mouratidis, K., 2002. The impact of financial liberalization policies on financial development: Evidence from developing economies. International Journal of Finance and Economics, 7, pp. 109–121.

Bai, J., Carvalho, D., Phillips, G., 2018. The impact of bank credit on labor reallocation and aggregate industry productivity. Journal of Finance, 73, pp. 2787–2836.

Bandiera, O., Caprio, J.G., Honohan, P., Schiantarelli, F., 2000. Does financial reform raise or reduce saving? Review of Economics and Statistics, 82, pp. 239–263.

Beccera, O., Cavallo, E., Scartascini, C., 2012. The politics of financial development: The role of interest groups and government capabilities. Journal of Banking and Finance, 36, pp. 626–643.

Beck, T., 2008. The Econometrics of Finance and Growth. The World Bank Policy Research Working Paper, No. 4608.

Beck, T., Degryse, H., Kneer, C., 2014. Is more finance better? Disentangling intermediation and size effects of financial systems. Journal of Financial Stability, 10, pp. 50–64.

Beck, T., Demirgüc-Kunt, A., Levine, R., 2003. Law, endowments and finance. Journal of Financial Economics, 70, pp. 137–181.

Beck, T., Levine, R., 2004. Stock markets, banks and growth: Panel evidence. Journal of Banking and Finance, 28, pp. 423–442.

Beck, T., Levine, R., Loayza, N., 2000. Finance and the sources of growth. Journal of Financial Economics, 58, pp. 261–300.

Bencivenga, V.R., Smith, B.D., 1991. Financial intermediation and endogenous growth. Review of Economics Studies, 58, pp. 195–209.

Benfratello, L., Schiantarelli, F., Sembenelli, A., 2008. Banks and innovation: Micro econometric evidence on Italian firms. Journal of Financial Economics, 90, pp. 197–217.

Berg, G., Schrader, J., 2012. Access to credit, natural disasters, and relationship lending. Journal of Financial Intermediation 21, pp. 549–568

Bernanke, B., Gertler, M., Gilchrist, S., 1999. The financial accelerator in a quantitative business cycle framework. In J. Taylor and M. Woodford (eds.), Handbook of Macroeconomics. Amsterdam: Elsevier, pp. 1341–1393.

Black, S.E., Strahan, P.E., 2002. Entrepreneurship and bank credit availability. Journal of Finance, 57, pp. 2807–2833.

Bolton, P., Despres, M., Pereira da Silva, L., Samama, F., and Svartzman, R. 2020a. The green swan: Central banking and financial stability in the age of climate change. Bank of International Settlements and Banque de France.

Bonnacorsi di Patti, E., Dell'Ariccia, G., 2004. Bank competition and firm creation. Journal of Money, Credit and Banking, 36, pp. 225–251.

Boot, A., 2000. Relationship banking: What do we know? Journal of Financial Intermediation, 9, pp. 7–25.

Bridges, J., Gregory, D., Nielsen, M., Pezzini, S., Radia, A., Spaltro, M., 2014. The impact of capital requirements on bank lending. Bank of England Working Paper, No. 486.

Brown, J.R., Gustafson, M.T., Ivanov, I.T., 2021. Weathering cash flow shocks. Journal of Finance, 76, pp. 1731–1772.

Brunnermeier, M., Sannikov, Y., 2014. A macroeconomic model with a financial sector. American Economic Review, 104, pp. 379–421.

Bruno, V., Hauswald, R., 2014. The real effect of foreign banks. Review of Finance, 18, pp. 1683–1716.

Campos, N.F., Karanasos, M.G., Tan, B., 2012. Two to tangle: Financial development, political instability and economic growth in Argentina. Journal of Banking and Finance, 36, 290–304.

Carrascosa, A., 2021. Should "brown" bank loans be penalized or "green" loans supported? Essade. Accessible at: https://dobetter.esade.edu/en/brown-green-loans.

Cecchetti, G., Kharroubi, E., 2013. Why does financial sector growth crowd out real economic growth? Finance and the Wealth of Nations Workshop, Federal Reserve Bank of San Francisco & The Institute of New Economic Thinking.

Celil, H.S., Oh, S., Selvam, S., 2022. Natural disasters and the role of regional lenders in economic recovery. Journal of Empirical Finance, 68, pp. 116–132.

Cetorelli, N., 2003. Life-cycle dynamics in industrial sectors. The role of banking market structure. Federal Reserve Bank of St. Louis Review, 85, pp. 135–147.

Cetorelli, N., Blank, M., 2019. Banking and real economic activity: Foregone conclusions and open challenges. In Allen N. Berger, Philip Molyneux, and John O. S. Wilson (eds.), The Oxford Handbook of Banking, 3rd edn. Oxford Handbooks, Oxford Academic.

Cetorelli, N., Gambera, M., 2001. Banking market structure, financial dependence and growth: International evidence from industry data. Journal of Finance, 56, 617–648.

Cetorelli, N., Strahan, P., 2006. Finance as a barrier to entry: Bank competition and industry structure in local US markets. Journal of Finance, 61, pp. 437–461.

Chavaz, M., 2016. Dis-integrating credit markets: Diversification, securitization, and lending in a recovery. Bank of England Unpublished Working Paper.

Claessens, S., Klingebiel, D., Laeven, L., 2005. Crisis resolution, policies, and institutions: Empirical evidence. In: Systemic Financial Crises: Containment and Resolution. Cambridge University Press, p. 169.

Claessens, S., Laeven, L., 2003. Financial development, property rights and growth. Journal of Finance, 58, pp. 2401–2436.

Claessens, S., Laeven, L., 2005. Financial dependence, banking sector competition, and economic growth. Journal of the European Economic Association, 3, pp. 179–220.

Collier, B.L., Babich, V.O., 2017. Financing recovery after disasters: Explaining community credit market responses to severe events. Journal of Risk and Insurance, 86, pp. 479–520.

Cornaggia, J., Mao, Y., Tian, X., Wolfe, B., 2015. Does banking competition affect innovation? Journal of Financial Economics, 115, pp. 189–209.

Cortés, K.R., 2014. Rebuilding after disaster strikes: How local lenders aid in the recovery. Working Paper. Bank of England, London.

Cortés, K.R., Strahan, P.E., 2017. Tracing out capital flows: How financially integrated banks respond to natural disasters. Journal of Financial Economics, 125, pp. 182–199.

Czura, K., Klooner, S., 2023. Financial market responses to a natural disaster: Evidence from credit networks and the Indian Ocean tsunami. Journal of Development Economics 160, 102996.

Dam, L., Kötter, M., 2012. Bank bailouts and moral hazard: Evidence from Germany. Review of Financial Studies, 25, pp. 2343–2380.

Dell'Ariccia, G., Detragiache, E., Rajan, R., 2008. The real effect of banking crises. Journal of Financial Intermediation, 17, pp. 89–112.

Demirgüc-Kunt, A., Horváth, B.L., Huizinga, H., 2017. How does long-term finance affect economic volatility? Journal of Financial Stability, 33, pp. 41–59.

Detragiache, E., Gupta, P., Tressel, T., 2008. Foreign banks in poor countries: Theory and evidence, Journal of Finance, 63, 2123–2160.

Dinger, V., Erman, L., te Kaat, D.M., 2022. Bank bailouts and economic growth: Evidence from cross-country, cross-industry data. Journal of Financial Stability 60, 100984.

Duanmu, J., Li, Y., Lin, M., Tahsin, S., 2022. Natural disaster risk and residential mortgage lending standards. Journal of Real Estate Research, 44, pp. 106–130.

Duqi, A., McGowan, D., Onali, E., Torluccio, G., 2021. Natural disasters and economic growth: The role of banking market structure. Journal of Corporate Finance, 71, 102101.

EBA, 2020. Annual Report. Accessible at: https://www.eba.europa.eu/sites/default/documents/files/document_library/About%20Us/Annual%20Repo rts/2020/1013723/EBA%202020%20Annual%20Report.pdf.

ECB, 2020. Guide on climate-related and environmental risks. Accessible at: https://www.bankingsupervision.europa.eu/ecb/pub/pdf/ssm.202011fin alguideonclimate-relatedandenvironmentalrisks~58213f6564.en.pdf.

Favara, G., Giannetti, M., 2015. Forced asset sales and the concentration of outstanding debt: Evidence from the mortgage market. CEPR Working Paper, 10476.

Fogel, K., Morck, R., Yeung, B., 2008. Big business stability and economic growth: Is what's good for General Motors good for America? Journal of Financial Economics, 89, pp. 83–108.

Gallagher, J., Hartley, D., 2017. Household finance after a natural disaster: The Case of Hurricane Katrina. American Economic Journal: Economic Policy, 9, pp. 199–228.

Gerschenkron, A., 1962. Economic Backwardness in Historical Perspective. Harvard University Press, Cambridge, MA.

Gertler, M., Karadi, P., 2015. Monetary policy surprises, credit costs, and economic activity. American Economic Journal: Macroeconomics, 7, pp. 44–76.

Gertler, M., Kiyotaki, N., 2010. Financial intermediation and credit policy in business cycle analysis. Handbook of Monetary Policy, 3, pp. 547–599.

Gertler, M., Kiyotaki, N., 2015. Banking, liquidity, and bank runs in an infinite horizon economy. American Economic Review, 105, pp. 2011–2043.

Giannetti, M., Ongena, S., 2009. Financial integration and firm performance: Evidence from foreign bank entry in emerging markets. Review of Finance, 13, pp. 181–223.

Gilje, E.P., Loutskina, E., Strahan, P.E., 2016. Exporting liquidity: Branch banking and financial integration. Journal of Finance, 71, pp. 1159–1184.

Gormley, T., 2008. The impact of foreign bank entry in emerging markets: Evidence from India. Journal of Financial Intermediation, 19, 26–51.

Hasan, I., Wachtel, P., Zhou, M., 2009. Institutional development, financial deepening and economic growth: Evidence from China. Journal of Banking and Finance, 33, pp. 157–170.

He, Z., Krishnamurthy, A., 2013. Intermediary asset pricing. American Economic Review, 103, pp. 732–770.

Henderson, D.J., Papageorgiou, C., Parmeter, C.F., 2013. Who benefits from financial development? New methods, new evidence. European Economic Review, 63, pp. 47–67.

Hosono, K., Miyakawa, D., Uchino, T., Hazama, M., Ono, A., Uchida, H., Uesugi, I., 2016. Natural disasters, damage to banks, and firm investment. International Economic Review, 57, pp. 1335–1370.

Hsu, P., Tian, X., Xu, Y., 2014. Financial development and innovation: Cross country evidence. Journal of Financial Economics, 112, pp. 116–135.

Huang, H.C., Lin, S.C., 2009. Non-linear finance–growth nexus. Economics of Transition, 17, pp. 439–466.

Imai, M., 2009. Political determinants of government loans in Japan. Journal of Law and Economics 52, 41–70.

IMF, 2019. Fiscal monitor. Accessible at: https://www.imf.org/en/Publications/FM/Issues/2019/10/16/Fiscal-Monitor-October-2019-How-to-Mitigate-Climate-Change-47027.

Inklaar, R., Koetter, M., Noth, F., 2015. Bank market power, factor reallocation, and aggregate growth. Journal of Financial Stability, 19, pp. 31–44.

Ivanov, I.T., Macchiavelli, M., Santos, J.A.C., 2022. Bank lending networks and the propagation of natural disasters. Financial Management, 51, pp. 903–927.

Jayaratne, J., Strahan, P.E., 1996. The Finance–growth nexus: Evidence from bank branch deregulation. Quarterly Journal of Economics, 111, pp. 639–670.

Jiang, F., Li, C.W., Qian, Y., 2019. Can firms run away from climate-change risk? Evidence from the pricing of bank loans. Retrieved from SSRN: https://ssrn.com/abstract=3477450.

Karakaya, N., Michalski, T.K., Örs, E., 2022. Banking integration and growth: Role of banks' previous industry exposure. Journal of Financial Intermediation, 49, 100944.

Kendall, J., 2012. Local financial development and growth. Journal of Banking and Finance, 36, pp. 1548–1562.

Kerr, W., Nanda, C., 2008. Democratizing entry: Banking deregulation, financing constraints and entrepreneurship. Journal of Financial Economics, 94, pp. 124–149.

Khwaja, A.I., Mian, A., 2005. Do lenders favor politically connected firms? Rent provision in an emerging financial market. Quarterly Journal of Economics, 120, pp. 1371–1411.

Khwaja, A.I, Mian, A., 2008. Tracing the impact of bank liquidity shocks: Evidence from an emerging market. American Economic Review, 98, pp. 1413–1442.

Kianersi, S., Jules, R., Zhang, Y., Luetke, M., Rosenberg, M., 2021. Associations between hurricane exposure, food insecurity, and microfinance; a cross-sectional study in Haiti. World Development, 145, 105530.

King, R.G., Levine, R., 1993. Finance and growth: Schumpeter might be right. Quarterly Journal of Economics, 108, pp. 717–738.

Koetter, M., Noth, F., Rehbein, O., 2020. Borrowers under water! Rare disasters, regional banks, and recovery lending. Journal of Financial Intermediation, 43, 100811.

Körner, T., Schnabel, I., 2009. Public ownership of banks and economic growth. Economics of Transition, 19, pp. 407–441.

La Porta, R., Lopez-de-Silanes, F., Shleifer, A., Vishny, R.W., 1998. Law and finance. Journal of Political Economy, 106, pp. 1113–1155.

La Porta, R., Lopez-de-Silanes, F., Shleifer, A., 2002. Government ownership of banks. The Journal of Finance, 57, 265–301.

Laeven, L., Valencia, F., 2013. The real effects of financial sector interventions during crises. Journal of Money Credit and Banking, 45, 147–177.

Law, S.H., Singh, N., 2014. Does too much finance harm economic growth? Journal of Banking and Finance 41, 36–44.

Levine, R., 1998. The legal environment, banks, and long-run economic growth. Journal of Money, Credit, and Banking, 30, pp. 596–620.

Levine, R., 1999. Law, finance, and economic growth. Journal of Financial Intermediation, 8, pp. 8–35.

Levine, O., Warusawitharana, M., 2021. Finance and productivity growth: Firm-level evidence. Journal of Monetary Economics, 117, pp. 91–107.

Levine, R., Zervos, S., 1998. Stock markets, banks, and economic growth. American Economic Review, 88, pp. 537–558.

Levine, R., Loayza, N., Beck, T., 2000. Financial intermediation and economic growth: Causes and causality. Journal of Monetary Economics, 46, pp. 31–77.

Loutskina, E., Strahan, P.E., 2011. Informed and uninformed investment in housing: The downside of diversification? Review of Financial Studies, 24, pp. 1447–1480.

Loutskina, E., Strahan, P.E., 2015. Financial integration, housing and economic volatility. Journal of Financial Economics, 115, pp. 25–41.

Morck, R., Yavuz, M.D., Yeung, B., 2011. Banking system control, capital allocation, and economy performance. Journal of Financial Economics, 100, pp. 264–283.

Nguyen, L., Wilson, J.O.S., 2020. How does credit supply react to a natural disaster? Evidence from the Indian Ocean Tsunami. European Journal of Finance, 26, pp. 802–819.

Odell, K.A., Weidenmeier, M.D., 2004. Real shock, monetary aftershock: The 1906 San Francisco earthquake and the panic of 1907. Journal of Economic History, 64, pp. 1002–1027.

Önder, Z., Özyıldırım, S., 2013. Role of bank credit on local growth: Do politics and crisis matter? Journal of Financial Stability 9, pp. 13– 25.

Ouzad, A., Kahn, M.E., 2022. Mortgage finance and climate change: Securitization dynamics in the aftermath of natural disasters. Review of Financial Studies, 35, pp. 3617–3665.

Peek, J., Rosengren, E.S., 2000. Collateral damage: Effects of the Japanese bank crisis on real activity in the United States. American Economic Review, 90, pp. 30–45.

Petersen, M.A., Rajan, R.G., 1995. The Effect of credit market competition on lending relationships. Quarterly Journal of Economics, 110, pp. 407–443.

Rajan, R.G., Zingales, L., 1998. Financial dependence and growth. American Economic Review, 88, pp. 559–586.

Rajan, R., Zingales, L., 2003. The great reversals: The politics of financial development in the twentieth century. Journal of Financial Economics, 69, pp. 5–50.

Rebhein, O., Ongena, S., 2022. Flooded through the back door: The role of bank capital in local shock spillovers. Journal of Financial and Quantitative Analysis, 57, pp. 2627–2658.

Rogers, T.M., 2012. Bank market structure and entrepreneurship. Small Business Economics, 39, pp. 909–920.

Romero-Ávila, D., 2007. Finance and growth in the EU: New evidence from the harmonisation of the banking industry. Journal of Banking and Finance 31, pp. 1937–1954.

Sapienza, P., 2004. The effects of government ownership on bank lending. Journal of Financial Economics, 72, pp. 357–384.

Schumpeter, J.A., 1911. The Theory of Economic Development. Harvard University Press, Cambridge, MA.

Schüwer, U., Lambert, C., Noth, F., 2019. How do banks react to catastrophic events? Evidence from hurricane Katrina. Review of Finance, 23, pp. 75–116.

Shen, C., Lee, C., 2006. Same financial development yet different economic growth—Why? Journal of Money, Credit, and Banking, 38, pp. 1907–1944.

Shleifer, A., Vishny, R., 1994. Politicians and firms. Quarterly Journal of Economics, 109, pp. 995–1025.

Stein, J., 1997. Internal capital markets and the competition for corporate resources. Journal of Finance, 52, pp. 111–133.

Stein, J., 2012. Monetary policy as financial stability regulation. Quarterly Journal of Economics, 127, pp. 57–95.

Thuy Nguyen, D.T., Diaz-Rainey, I., Roberts, H., Le, M., 2023. The impact of natural disasters on bank performance and the moderating role of financial integration. Applied Economics. https://doi.org/10.1080/00036846.2023.2174931.

Von Pischke, J., 2002. Innovation in finance and movement to client-centered credit. Journal of International Development, 14, pp. 369–380.

CHAPTER 5

The Role of Other Actors in Promoting Post-Disaster Economic Recovery in Partnership with Banks

Abstract The international community is constantly bringing to the public attention that disasters are becoming more expensive and are increasing poverty, especially in low and low- middle-income countries. The adaptation finance gap in developing countries is five to 10 times greater than the current international financial aid and continues to widen. Multilateral agencies such as the World Bank have launched several programs aiming at helping countries understand, manage, and reduce natural hazard risks. Various risk transfer mechanisms, such as insurance tools, are available and have proven successful. Many countries have introduced climate strategies and policies. Governments need to actively establish a risk reduction culture, which should incentivize investment in disaster resilience, understand climate projections by working closely with the scientific community, and target humanitarian assistance to the most vulnerable through grants or subsidized loans.

Keywords Disaster relief · Disaster risk management and reduction · Adaptation and mitigation finance · Climate risk retention and transfer instruments

© The Author(s), under exclusive license to Springer Nature Switzerland AG 2023
A. Duqi, *Banking Institutions and Natural Disasters*,
Palgrave Studies in Impact Finance,
https://doi.org/10.1007/978-3-031-36371-9_5

5.1   SUPRA-NATIONAL INSTITUTIONS
AND DEVELOPMENT BANKS

The international community considers natural disasters and weather hazards as part of the consequences of climate change since it has been acknowledged that their increased frequency and magnitude are directly related to the rising temperatures on our planet. Therefore, actions to prevent significant damages from natural disasters and to promote post-disaster economic recovery are intrinsically related to actions against climate change which should lead to a sustainable future for all countries. At the same time, international organizations have established agencies that look specifically at disaster relief, especially for less developed nations.

In this section, the role of the most important international bodies leading the fight against climate change and those supporting countries in building resilience against weather hazards will be examined with a specific focus on the role of finance in this context. It is worth noting that the role of bank regulators is not discussed here but in Chapters 3 and 4.

5.1.1   United Nations Framework Convention on Climate Change (UNFCCC)

The United Nations (U.N.) has long been the promoter of international cooperation since its establishment in 1945. The U.N. system comprises several funds, programs, and specialized agencies with their own work area, leadership, and budget. Several agencies are involved in actions against climate change and management/adaptation to disaster risks. Their actions are based on the United Nations Framework Convention on Climate Change (UNFCCC), an international environmental treaty aimed at combating dangerous human interference with the climate system. It was signed in Rio de Janeiro in June 1992. The ultimate objective of the Framework Convention is "the stabilization of greenhouse gas concentrations in the atmosphere at a level that would prevent dangerous anthropogenic (i.e., human-caused) interference with the climate system.[1]

The UNFCCC was opened for signature on May 9, 1992, after an Intergovernmental Negotiating Committee produced the text of the

---

[1] Discussing the goals and actions of UNFCCC is beyond the scope of this study. Interested readers can find a detailed description of UNFCCC activities and mission here.

Framework Convention as a report following its meeting in New York. It entered into force in March 1994. Countries that sign up for the UNFCCC are known as "Parties." Currently, there are 192 signed-up Parties.

Since the UNFCCC entered into force, the Parties have been meeting annually in Conferences of the Parties (COP) to assess progress in dealing with climate change and beginning in the mid-1990s, to negotiate the Kyoto Protocol, which established legally binding obligations for developed countries to reduce their greenhouse gas emissions.

In 2021, during COP21, finance was extensively discussed, and consensus was reached on the need to continue increasing support to developing countries. Developed country Parties reaffirmed their duty to fulfill the pledge of providing USD 100 billion annually by 2023 at the latest. Parties agreed on a way forward concerning the post-2025 climate finance goal.

5.1.2 United Nations Office for Disaster Risk Reduction (UNDRR)

The principal body of the U.N. whose mission is to oversee programs for disaster risk reduction, is the UNDRR, the United Nations Office for Disaster Risk Reduction. The mission of the UNDRR is to help decision-makers across the globe better understand and change their attitude to risk. The response of many countries as they try to cope with natural disasters suffers from a bias related to only focusing on post-disaster actions to provide relief to communities that have suffered losses from these hazards. It is generally overlooked that natural hazards cause damage because of human errors in how they live, build, plan, and invest. If careful and coordinated planning is implemented, people's exposure and vulnerability to natural hazards will be significantly curbed. Hazards are traditionally seen as the result of external processes, that is, outside human influence in their probability of occurrence, magnitude, intensity, speed, velocity, and persistence. In climatic events, though, the mounting evidence is that human intervention can alter hazards, as is the case of the changing rainfall and decreased humidity associated with deforestation; "heat islands" as a consequence of urban concentrations; and the alteration of seasonality in terms of drought and extreme precipitation (Zapata-Marti 2013).

Disasters' consequences are becoming more severe than before because they are related to human activity on the planet. Examples could be the exponential growth of the world population, its concentration in megacities such as Tokyo, Cairo, New Mexico, or Lagos, or human settlements in very exposed regions such as sea costs or near river basins (Munich Re Group 1999).

Since we are witnessing an increase in the frequency, force, and characteristics of natural disasters, their cumulative impact reduces the capacity of households and authorities to cope and the resilience of societies and human activities. In the face of this, risk reduction requires not only mitigating measures for historical patterns of hazards but also adaptation to new climatic trends and changes.

Since 2015 the UNDRR has developed the Sendai Framework for Disaster Risk Reduction 2015–2030. It provides U.N. member states with concrete actions to protect development gains from disaster risk. The Sendai Framework is closely related to other 2030 Agenda agreements such as the Paris Agreement on Climate Change, The Addis Ababa Action Agenda on Financing for Development, the New Urban Agenda, and ultimately the Sustainable Development Goals.

Figure 5.1 outlines the scope and purpose of the Sendai Framework, its expected outcomes, goals, and targets. The main goal of the Framework is to implement integrated and inclusive measures that prevent and reduce hazard exposure and vulnerability to disasters, increase preparedness for response and recovery, and thus strengthen resilience. One of the main concerns of UNDRR is to improve the capacity of less developed countries, including the mobilization of support through international cooperation.

Several guiding principles and priorities aim to understand disaster risk, strengthen disaster risk governance to manage disaster risk, invest in disaster risk reduction for resilience, enhance disaster preparedness for effective response, and "Build Back Better" in recovery, rehabilitation, and reconstruction.

The role of finance is related to priority 3, which points to a better allocation of resources for developing and implementing disaster risk reduction strategies, policies, plans, laws, and regulations in all relevant sectors. Moreover, mechanisms for disaster risk transfer, risk-sharing and retention, and financial protection should be promoted.

The Framework continuously brings to the public attention the need to specifically focus on developing countries more exposed to disasters, such as small islands developing states, landlocked developing countries, African states, and middle-income countries facing specific challenges.

Chart of the Sendai Framework for Disaster Risk Reduction
2015–2030

Scope and purpose

The present framework will apply to the risk of small-scale and large-scale, frequent and infrequent, sudden and slow-onset disasters, caused by natural or manmade hazards as well as related environmental, technological and biological hazards and risks. It aims to guide the multi-hazard management of disaster risk in development at all levels as well as within and across all sectors

Expected outcome

The substantial reduction of disaster risk and losses in lives, livelihoods and health and in the economic, physical, social, cultural and environmental assets of persons, businesses, communities and countries

Goal

Prevent new and reduce existing disaster risk through the implementation of integrated and inclusive economic, structural, legal, social, health, cultural, educational, environmental, technological, political and institutional measures that prevent and reduce hazard exposure and vulnerability to disaster, increase preparedness for response and recovery, and thus strengthen resilience

Targets

Substantially reduce global disaster mortality by 2030, aiming to lower average per 100,000 global mortality between 2020–2030 compared to 2005–2015	Substantially reduce the number of affected people globally by 2030, aiming to lower the average global figure per 100,000 between 2020–2030 compared to 2005–2015	Reduce direct disaster economic loss in relation to global gross domestic product (GDP) by 2030	Substantially reduce disaster damage to critical infrastructure and disruption of basic services, among them health and educational facilities, including through developing their resilience by 2030	Substantially increase the number of countries with national and local disaster risk reduction strategies by 2020	Substantially enhance international cooperation to developing countries through adequate and sustainable support to complement their national actions for implementation of this framework by 2030	Substantially increase the availability of and access to multi-hazard early warning systems and disaster risk information and assessments to people by 2030

Priorities for Action

There is a need for focused action within and across sectors by States at local, national, regional and global levels in the following four priority areas.

Priority 1 Understanding disaster risk	Priority 2 Strengthening disaster risk governance to manage disaster risk	Priority 3 Investing in disaster risk reduction for resilience	Priority 4 Enhancing disaster preparedness for effective response, and to «Build Back Better» in recovery, rehabilitation and reconstruction
Disaster risk management needs to be based on an understanding of disaster risk in all its dimensions of vulnerability, capacity, exposure of persons and assets, hazard characteristics and the environment	Disaster risk governance at the national, regional and global levels is vital to the management of disaster risk reduction in all sectors and ensuring the coherence of national and local frameworks of laws, regulations and public policies that, by defining roles and responsibilities, guide, encourage and incentivize the public and private sectors to take action and address disaster risk	Public and private investment in disaster risk prevention and reduction through structural and non-structural measures are essential to enhance the economic, social, health and cultural resilience of persons, communities, countries and their assets, as well as the environment. These can be drivers of innovation, growth and job creation. Such measures are cost-effective and instrumental to save lives, prevent and reduce losses and ensure effective recovery and rehabilitation	Experience indicates that disaster preparedness needs to be strengthened for more effective response and ensure capacities are in place for effective recovery. Disasters have also demonstrated that the recovery, rehabilitation and reconstruction phase, which needs to be prepared ahead of the disaster, is an opportunity to «Build Back Better» through integrating disaster risk reduction measures. Women and persons with disabilities should publicly lead and promote gender-equitable and universally accessible approaches during the response and reconstruction phases

Guiding Principles

Primary responsibility of States to prevent and reduce disaster risk, including through cooperation	Shared responsibility between central Government and national authorities, sectors and stakeholders as appropriate to national circumstances	Protection of persons and their assets while promoting and protecting all human rights including the right to development	Engagement from all of society	Full engagement of all State institutions of an executive and legislative nature at national and local levels	Empowerment of local authorities and communities through resources, incentives and decision-making responsibilities as appropriate	Decision-making to be inclusive and risk-informed while using a multi-hazard approach
Coherence of disaster risk reduction and sustainable development policies, plans, practices and mechanisms, across different sectors	Accounting of local and specific characteristics of disaster risks when determining measures to reduce risk	Addressing underlying risk factors cost-effectively through investment versus relying primarily on post-disaster response and recovery	«Build Back Better» for preventing the creation of and reducing existing, disaster risk	The quality of global partnership and international cooperation to be effective, meaningful and strong	Support from developed countries and partners to developing countries to be tailored according to needs and priorities as identified by them	

Fig. 5.1 Sendai framework for disaster risk reduction (*Source* UNDRR GAR 2022)

Developing countries need enhanced international support for disaster risk reduction and to receive financial support and loans from multilateral organizations such as the World Bank.

The UNDRR releases every year a Global Assessment Report on Disaster Risk Reduction. The latest report published in 2022 reviews the progress toward the achievement of the Sendai Framework goals set for 2030.[2] Some of the main findings of the report are:

- Disasters are claiming fewer lives, but costing more and are increasing poverty. Low-income and lower-middle-income countries lose on average 0.8–1% of their national GDP to disasters annually, compared to 0.1% and 0.3% in high-income and upper-middle-income countries, respectively.
- Less than half of disaster-related losses at a global level in 2020 were insured, and insurance is overwhelmingly concentrated in wealthier countries.
- Early analysis of the data reported by Member States through SFM indicates the global community will miss achieving the Sendai Framework goals by 2030.
- Most resources have supported activities to respond to and recover from disasters, although disaster-related financing has increased since 2010 (Fig. 5.2).
- Only a negligible share of funding for DRR, per capita, goes to countries with the highest disaster-related mortality rates.
- Some countries with the highest Natural Hazard Risk Index receive commensurate levels of prevention and preparedness funding, while most do not (Fig. 5.3).

Some important takeaways of the report are related to the need to treat and manage disaster risk similarly to systemic risk for banks. One of the following three approaches could be adopted:

- Reduce vulnerabilities and weak nodes in the system (i.e., construct dams to protect system-relevant power plants) and minimize cascading effects.

---

[2] The full report is available at this weblink.

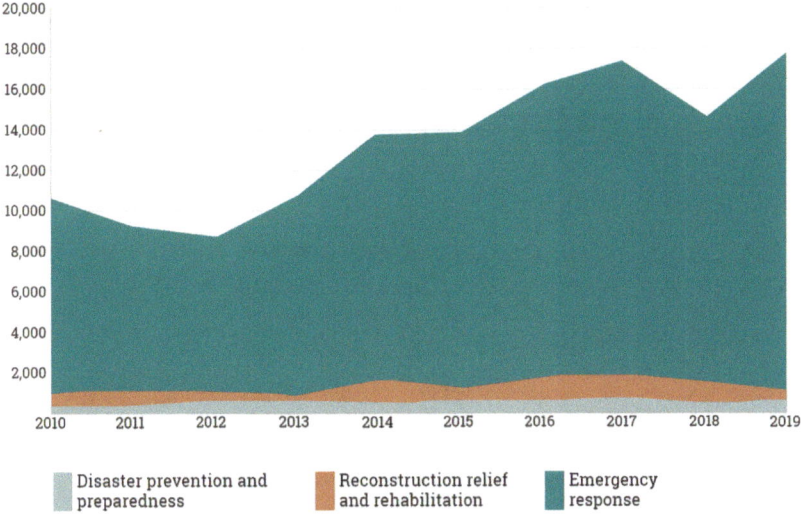

Fig. 5.2 Disaster-related financing: Official Development Assistance (ODA) for prevention and preparedness, funding for reconstruction relief and rehabilitation, and emergency response ($ million), 2010–2019 (*Source* UNDRR GAR 2022)

- Implement strategies to avoid risk propagation by identifying interdependencies.
- Adopt strategies to change agent behavior, such as reducing the propensity toward systemically risky behavior (i.e., introducing a systemic risk tax).

Countries' strategies to cope with natural disasters and the identification of vulnerabilities or interdependencies depend on the quality and granularity of data. The intersection of geospatial databases of infrastructure assets and hazard maps provides an indication of the likelihood that a particular asset will be affected compared to others. Adshead et al. (2020) studied how a small island like Curacao would cope with the consequences of a 1-meter sea level rise and a 4-meter storm surge event.

They found that exposure was predominantly concentrated in the capital city with potential interdependent consequences for the service infrastructure in that area. Fuldauer et al. (2021), in their study about the effects of tropical storms in Saint Lucia, show that a catastrophic

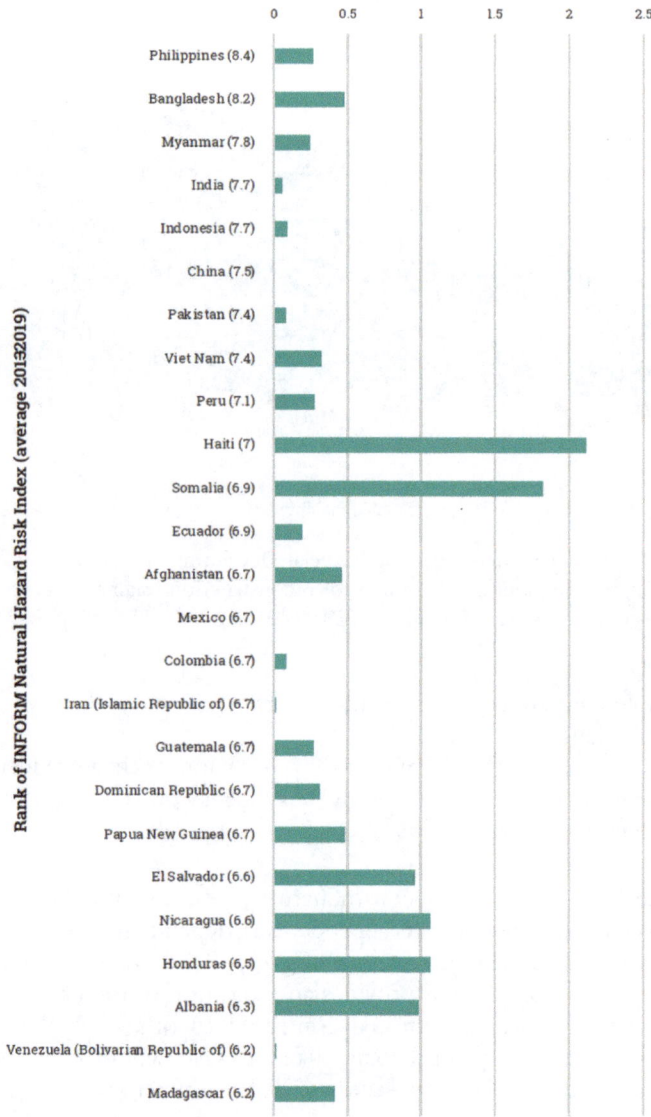

Fig. 5.3 ODA for prevention and preparedness received by countries with the highest level of Natural Hazard Risk, 2010–2019 average values (*Source* UNDRR GAR 2022)

event could generate massive losses in the freight sector, impacting the import of goods and services vital for the economy of the island. Therefore, investing in the risk reduction capacity of the country's two main ports could increase the island's resilience to natural disasters (Pant et al. 2022).

Since the modeling of the effects of climate change on society and the economy are rather complex, recently, other methodologies are attempting to represent all the possible interactions by "climate storylines." A project funded by the European Union, namely Remote Climate Effects and their Impact on European Sustainability, Policy and Trade (RECEIPT Project), provides an example of a "climate storyline." The researchers studied the cocoa market, focusing on cocoa producers in Western Africa. The goal was to understand how different actors in this industry perceive climate change, what kind of data they deem essential for the sector, and what type of information and analysis is needed to react to extreme weather events induced by climate change.

The cocoa producers in Western Africa were mainly concerned by erratic rainfall patterns as critical climate hazards that led to uncertainty about planting times, reduced quality of cocoa beans, or crop losses. Possible risk reduction and adaptation strategies included installing irrigation systems, reforestation, or soil water conservation (UNDRR GAR 2022). The producers faced high opportunity costs for moving out of cocoa production as cocoa plants remained fertile for more than 60 years. Moreover, simplified risk models had not considered the cultural importance of cocoa in some communities, which is seen as a status symbol. Therefore, farmers might be reluctant to move to other crops, even when production is no longer economically viable. Several interdependencies were discovered among variables such as climate shocks, pest, and crop management, various diseases, shifting production to alternative crops, the chocolate industry, etc. Figure 5.4 presents key elements of this storyline. Successively, researchers could estimate the impact of climate risks on cocoa supply using a range of qualitative–quantitative models incorporating systemic risk in the analysis.

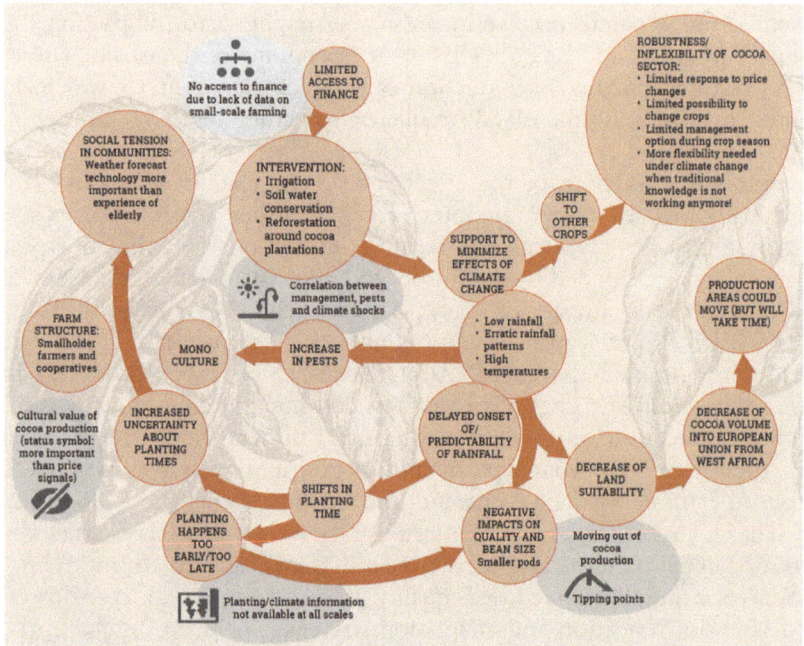

Fig. 5.4 Cocoa storyline: Climate change impact in Western Africa (*Source* The UNDRR GAR 2022)

5.1.3   United Nations Environment Programme (UNEP) and the United Nations Environment Programme Finance Initiative (UNEP FI)

UNEP is the leading agency inside the U.N. system, which advocates environmental policies, sets the environmental agenda, and promotes the coherent implementation of the environmental dimension of sustainable development within the U.N. system. UNEP reviews the advancement toward the objective of the Paris Agreement and publishes its main conclusions in its annual reports. The reports highlight the need to implement adaptation and mitigation measures related to climate change. Adaptation finance represents the funding of all investments that should reduce the risks and harm communities might suffer from climate hazards. Mitigation finance, instead, focuses on financing projects aimed at reducing GHGs, such as increasing renewable energy capacity. The latest UNEP Global Adaptation Report evidences that[3]:

- Climate risks are increasing as global warming accelerates. Substantial mitigation and adaptation are both keys to avoiding hard adaptation limits.
- Global efforts in adaptation planning, financing, and implementation continue to make incremental progress but need to catch up with increasing climate risks.
- The adaptation finance gap in developing countries is likely five to 10 times greater than current international adaptation finance flows and continues to widen.
- Considering interlinkages of adaptation and mitigation action from the outset in planning, finance, and implementation can enhance co-benefits.

The adaptation finance needs in developing countries are estimated at an average of USD 202 billion annually (Table 5.1 provides a breakdown by geographic area). Although the trend of international financial support toward these countries has gradually increased, most of the funds are allocated to mitigation efforts. A growing body of research shows that finance providers are not strategically targeting adaptation assistance toward the most vulnerable countries and population groups.

Figure 5.5 provides a breakdown of mitigation and adaptation finance provided by several multilateral development banks (MDBs) as of 2021. It confirms that mitigation finance exceeds adaptation finance for each institution.

The U.N. and its partners have launched in recent years several funds aiming at adaptation actions across the globe, namely the Adaptation Fund (AF), and the Green Climate Fund (GCF). There are two climate funds under the Global Environmental Facility (GEF), the Least Developed Countries Fund (LDCF) and the Special Climate Change Fund (SCCF), which are of particular relevance to developing countries. These facilities funded more than 470 projects between 2006 and 2022 addressing mainly drought, floods and rainfall variability. However, they

---

[3] The UNEP 2022 Adaptation Report can be accessed at this weblink.

Table 5.1 Potential developing countries' adaptation finance needs for 2021–2030 by region

Region	Median annual adaptation finance needs in USD billion (2020 value)	Annual adaptation finance needs as a percentage of GDP
East Asia and Pacific	69	0.35
South Asia	59	1.69
Sub-Saharan Africa	36	2.10
Latin America and Caribbean	21	0.41
Middle East and North Africa	15	0.47
Europe and Central Asia	4	0.69
Global	**202**	**0.60**

Source UNEP (2022)

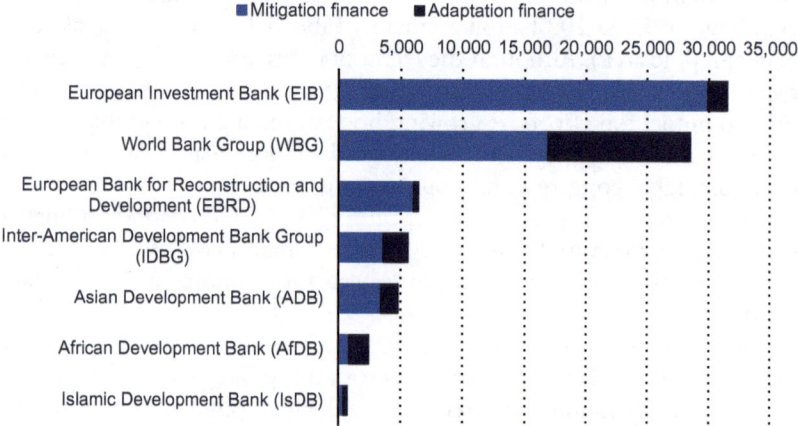

Fig. 5.5 Climate finance from MDBs in 2021 (by activity in USD million) (*Source* AfDB et al. 2022 and Statista database)

represent only a tiny fraction of the multilateral adaptation funds allocated in 2020. At the same time, developing countries also use domestic sources to fund adaptation efforts, although the extent varies with the level of economic development, political priorities, and other factors.

The UNEP FI is a network of financial institutions partnered with the U.N. and other organizations to shape the sustainable finance agenda. It supports financial institutions in setting and implementing sustainability targets and global frameworks developed by leading practitioners, conducting in-depth thematic research, guidance, and communities of practice. Therefore UNEP FI acts as an advisory body by supporting financial institutions across the globe on sustainability concerns. Among other things, UNEP FI launched the Task Force on Nature-Related Financial Disclosure (TNFD) which fosters market transparency, aiming at delivering a risk management and disclosure framework for organizations to report and act on nature-related risks. Recently, it introduced an online risk tool to assist financial institutions in visualizing how their portfolio depends on and impacts nature and how environmental change can create risks for the businesses they finance. UNEP FI has launched several networks among banks, such as Principles of Responsible Banking and the Net Zero Banking Alliance (NZBA). Through the Principles, banks take action to align their core strategy, decision-making, lending, and investment with the U.N. Sustainable Development Goals and international agreements such as the Paris Climate Agreement. The NZBA is the climate-focused initiative of this global Framework.

5.1.4   Multilateral Development Banks

Multilateral development banks (MDBs) are supranational institutions set up by sovereign states. Their remits reflect the development aid and cooperation policies established by these states. They have the common task of fostering economic and social progress in developing countries by financing projects, supporting investment, and generating capital to benefit all global citizens.

MDBs operate in a few key strategic areas, such as climate action, jobs, growth, or migration. Often they co-participate in various projects of a large entity. Currently, the most important MDBs are the World Bank Group, the European Investment Bank, the European Bank for Reconstruction and Development, the African Development Bank, the Asian Development Bank, the Asian Infrastructure Development Bank, the Inter-American Development Bank, and the Islamic Development Bank. In 2021 these MDBs committed a total of USD 50.666 billion of climate finance to low and middle-income economies and USD 31.051 billion to high-income economies. Figure 5.6 presents a breakdown of these numbers by single MDB. Figure 5.7 highlights the total climate finance provided by MDB divided by activity (mitigation vs. adaptation finance).

The following will focus on the World Bank Group, the leading global MDB per volume of activity committed to climate finance. However, its mission and modus operandi is similar to the other MDBs, except that the latter has a regional focus on the projects they finance.

The World Bank Group, established in 1947, provides a wide array of financial products and technical assistance and helps countries share and apply innovative knowledge and solutions to the challenges they face. The WBG is the largest multilateral provider of climate finance for developing countries and has increased financing to record levels over the past two years. Currently, the institution is implementing its "Climate Change Action Plan 2021–2025," which pursues poverty eradication and shared prosperity with a sustainability lens.

It aims to build a robust analytical base at the country level by introducing Country Climate and Development Reports (CCDRs) that address the interplay between climate and development.

The WBG supports countries to design and implement macroeconomic policies by developing climate-centered macro-projections and to assess the impact of climate shocks in terms of physical risks on macroeconomic

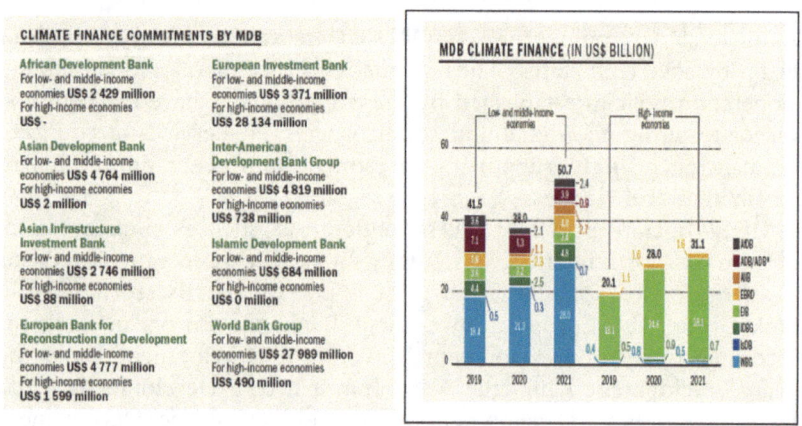

Fig. 5.6 Climate finance breakdown by single MDB (*Source* EIB 2021)

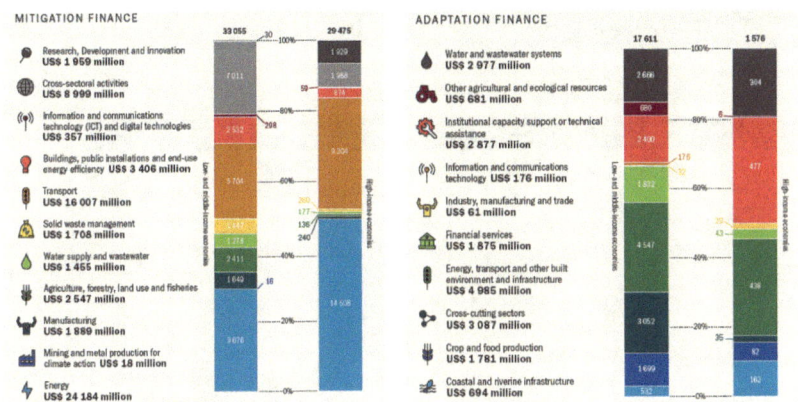

Fig. 5.7 Total MDBs' climate finance by activity in USD million (*Source* EIB 2021)

outcomes and fiscal sustainability. Moreover, it fosters the introduction of climate policies in country macroeconomic frameworks and the integration of climate planning into national budgets and expenditure frameworks. Overall, the institution aims to support the "greening" of the economies rather than financing specific "green" projects.

The WBG is committing to achieving 35% in climate finance for the entire group. Figure 5.6 evidenced that almost half of the financing provided by the WBG is devoted to adaptation projects. This will continue to be the case even in the future. It should increase the resilience of less developed and emerging countries to natural hazards and reduce their vulnerability. For example, the WBG will step up support for climate-smart agriculture (CSA) across the entire agriculture and food value chains, which will help countries manage flood and drought risks together, reducing the water-related shocks and protecting livelihoods and productive resources.

The WBG has launched several programs aiming at helping countries, especially low and middle-income ones, understand, manage, and reduce their risks from natural hazards and climate change. One of them, the Global Facility for Disaster Reduction and Recovery (GFDRR), established in 2006, provides granting and expertise for policy advice on improving disaster risk management (DRM) at national and local levels. Figure 5.8 highlights the GFDRR Logical Framework.

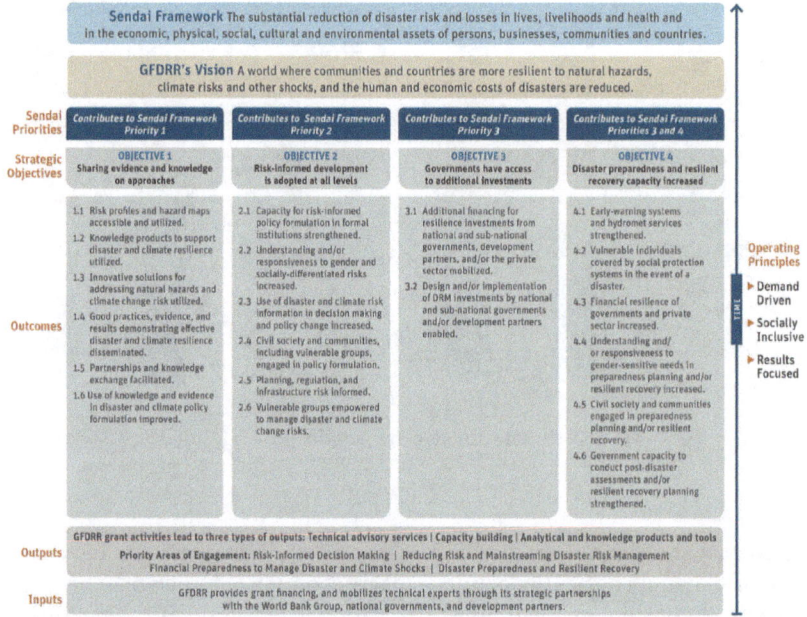

Fig. 5.8 The GFDRR logical framework (*Source* GFDRR 2022 weblink)

In 2010 the WBG established another program, the Disaster Risk Financing and Insurance Program (DRFIP), which aims at helping countries develop comprehensive financial protection strategies. DRFIP has supported governments, businesses, and households in more than 60 countries by bringing together the World Bank Group's financial, analytical, advisory, and convening services.

Recently, the GFDRR, in partnership with the Governments of the United Kingdom and Germany, established the Global Risk Financing Facility (GRiF) to strengthen the resilience of the most vulnerable countries to climate and disaster shocks. GRiF project grants support the development and implementation of risk retention and risk transfer instrument such as:

- Contingency funds
- Sovereign insurance
- Domestic insurance

- Risk pools
- Partial portfolio credit guarantee schemes.

Below, two examples of GRiF projects are provided. The first one illustrates how GRiF implements a pooling scheme in Indonesia (Box 5.1). This financial strategy combines the risks a group of countries faces into a single portfolio so that each contributor's share of the portfolio is less risky than its individual share. Pooling schemes bring additional benefits to the countries that decide to join, such as establishing common institutional structures, planning for contingencies, investing in reliable risk data, and agreeing to structured monitoring and evaluation.

The second example highlights how GRiF enhances the financial resilience of Rwanda's micro, small, and medium enterprises (MSMEs) (Box 5.2). MSMEs are especially vulnerable to disaster shocks in developing countries because of financial constraints due to size and information asymmetries. GRiF offers partial portfolio guarantees that encourage local banks to provide more lending, ensuring that MSMEs can access credit and contribute to the economy.

Box. 5.1: I N D O N E S I A: Pooling Funds Boosts Transparency and Effectiveness of the National Disaster Risk Strategy

As part of the so-called Pacific Ring of Fire, the islands of Indonesia are subject to more volcanic and earthquake activity than anywhere else on Earth. Managing disaster risks is especially critical along this path, where 90% of the world's earthquakes occur. To address these risks, in 2018, the Indonesian Government partnered with the World Bank to launch a disaster risk finance and insurance strategy. Until then, the country relied on its state budget to finance disaster recovery efforts. The strategy aimed to bring more predictable financing to disaster preparedness and relief efforts and reduce the likelihood that funds earmarked for programs such as health and education programs get reallocated in years of severe disasters or climate events.

Leveraging lessons learned from around the world to inform Indonesia's new pooling fund

Indonesia established a pooling fund as part of a $500 million World Bank lending operation. The lending operation supported Indonesia's efforts to strengthen its financial response to natural disasters, climate risks, and health-related shocks such as the COVID-19 pandemic. By setting up the Pooling Fund for Disasters (Pooling Fund *untuk Bencana*), the Indonesian Government would have a centralized mechanism that improves planning and manages the flow of disaster-related funding among government agencies to disaster victims. The pooling fund should not only increase efficiency, but also would promote greater transparency about how funds are used.

GRiF provides critical seed funding

Start-up and operating costs to establish the pooling fund as an institution were financed through a $14 million GRiF grant. The grant further supported the technical groundwork needed to identify and develop appropriate risk transfer solutions to backstop the pooling fund in years of severe disasters or climate events. The pooling fund has been operational since 2022 and can become a model of risk financing expertise at regional and international levels. Work is underway to integrate Indonesia's efforts with another GRiF-supported initiative, SEADRIF, a platform for Association of Southeast Asian Nations (ASEAN) countries to access disaster risk financing solutions. As a final component, GRiF funding would help expand the 2019 pilot of the Government's state asset insurance program. The program transferred public asset risk to a consortium of more than 50 local insurance and reinsurance companies. These institutions manages to obtain further reinsurance in international markets. By the end of 2022, all government ministries should be covered by the program, which could later be scaled up to cover other infrastructure assets.

Source GRiF (2021) Annual Report

Box 5.2: Bridge Lending Window to Support an Estimated 2000 MSMEs in Rwanda in 2021

MSMEs represent 98% of businesses and contribute 55% of the total GDP of Rwanda. Although they are essential to the economy, one-third of Rwandan MSMEs stated in a 2019 World Bank survey that access to finance was their single biggest business constraint—a more significant share than any other cited constraint. The credit crunch is particularly pronounced in agro-based businesses, where credit extended to the agriculture sector accounts for less than 2% of all lending.

How GRiF provided support

GRiF awarded a grant of $8.5 million so that banks could continue lending to MSMEs in the agricultural sector struggling to recover from drought or the COVID-19 pandemic. The grant capitalized a bridge lending window, covered insurance premiums, and provided technical support to enhance disaster risk financing capacity in the government and private sector. The bridge lending window aimed to reach 2000 MSMEs, 60% owned by women.

Progress in 2021

During 2021, the project team supported the Development Bank of Rwanda in creating a project implementation unit to coordinate the project over the next five years. The next steps include determining which MSMEs need short-term bridge loans, carrying out requisite due diligence, developing the insurance backstop product to protect the bridge lending window's capital from depletion, and collaborating with financial institutions using the bridge lending window to extend credit to MSMEs.

Source GRiF 2021 Annual Report

5.1.5   *The International Monetary Fund*

The International Monetary Fund (IMF) advances international monetary cooperation, promotes the expansion of trade and growth, and discourages policies that would harm prosperity. It monitors the economic and financial performance of countries through policy advice, provides financial assistance to member countries, and assists governments to implement sound economic policies by offering technical assistance and training. The IMF is offering policy advice on how countries can adapt and mitigate climate change as this represents a significant threat to long-term growth and prosperity. It carries out and releases research on the economic implications of climate change to its members to help them capture sustainable growth opportunities. The IMF acknowledges that climate change will have a significant impact on many countries especially low-income ones, therefore, their macroeconomic policies need to be adapted to accommodate more weather shocks such as severe natural disasters.

Since 2017, the IMF introduced the Climate Change Policy Assessments (CCPA) in partnership with the World Bank. Through the CCPA, a thorough assessment of countries' climate strategies is provided to assist them in building robust macro-frameworks for responding to climate change and improving the prospects of attracting external finance. Since

2017, CCPA pilots have been completed for Seychelles, St Lucia, and Belize; a CCPA is in progress for Grenada and planned for Micronesia and Tonga.

In 2019 the IMF published a report providing guidelines for low-income countries to improve their resilience to natural disasters. This strategy should be based on three pillars, improving resilient infrastructure, creating fiscal buffers, and using pre-arranged financial instruments to protect fiscal sustainability and manage recovery costs, contingency planning, and related investments, ensuring a speedy response to a disaster.[4]

The IMF's role in building resilience is based on its experience in analyzing and advising on macro-critical issues and supporting associated capacity development. The IMF assists and integrates resilience financing into national macro-fiscal strategies and evaluates the relative sustainability. Small countries could face difficulties financing their debt since the adaptation costs are incurred upfront, but the benefits will eventually arrive only in the long term. As a consequence, international investors could find the debt dynamics not sustainable. The IMF, by providing a sound macroeconomic analysis of debt sustainability in the long term, can assure markets about the countries' ability to honor their financial obligations.

The IMF has accumulated a long experience evaluating the impact of disaster risks on the macroeconomic outlook. IMF researchers employ various models which reflect data limitations and country-specific features. Box 5.3 describes an innovative model of integrating the cost of natural disasters in the Pacific islands. The shock that natural disasters bring into the countries' public debt can be incorporated into a baseline scenario to assess the sustainability of debt in the medium-long term (*tail shock*). The analysis can be enriched by including the costs and benefits of insurance under various disaster shocks and potential risk transfer instruments. Moreover, the models can comprise the impact of natural disasters on countries' external balance and real exchange rates. It emerges that resilience investments can assure a better sustainability of debt, which has been previously overlooked.

---

[4] IMF Policy Paper No. 2019/020.

The IMF provides a range of funding options to support the implementation of resilience strategies. These include:

- A disbursing arrangement for countries facing balance of payments needs in implementing their resilience-focused medium-term programs.
- A precautionary three-year stand-by arrangement, for countries with potential Balance of Payments (BoP) needs as insurance against an adverse shock (whether disaster-related or other) while implementing their Fund-supported program.
- A non-financial signaling instrument (the Policy Coordination or Policy Support Instruments) for countries seeking to signal IMF endorsement of their economic program, which could facilitate access to Fund resources in case of a BoP need from an adverse shock
- Post-disaster financial assistance via the Rapid Financing Instrument (RFI) or Rapid Credit Facility (RCF) to assist countries with an urgent BoP need when hit by adverse exogenous shocks such as a natural disaster.

Regarding the support aiming at capacity development, the IMF can strengthen countries' capacity to fund and manage large-scale infrastructural investment programs. These could be incorporated into medium-term fiscal and budgeting frameworks. A Medium Term Revenue Strategy could strengthen the mobilization of domestic revenues by acting on tax policy frameworks, legislation, and revenue administration. Moreover, by building robust Public Financial Management systems, the IMF can ensure that Public Expenditure and Financial Accountability (PEFA) assessments include measures that facilitate access to Climate Funds.

Regarding financial resilience, the IMF can assist countries in building an adequate financial infrastructure and develop an understanding of disaster risks and main risk transfer instruments. The IMF, together with the World Bank can use moral suasion powers to coordinate various stakeholders, such as private insurers, governments, regional pools, donors, and climate funds, to resolve existing hurdles to accessing market-based risk transfer mechanisms, including exploring the financial viability of debt instruments with disaster clauses and addressing scale obstacles to the development of insurance products.

Regarding post-disaster response, the IMF can promote business continuity plans for both the central bank and commercial banks by assessing the resilience of banks' loan portfolios to disaster shocks.[5]

Box 5.3: Integrating the Cost of Natural Disasters in the Pacific Islands

IMF staff investigated the impact of natural disasters in Pacific Island Countries (Lee et al. 2018). The paper highlights the intensity of natural disasters for each country in the Pacific based on the distribution of damage and population affected by disasters. It estimates the impact of disasters on economic growth and international trade using a panel regression. The results show that severe disasters have significantly and adversely impact on economic growth and lead to a deterioration of the fiscal and trade balances. The paper then identifies a simple and consistent method to adjust staff's economic projections and debt sustainability analysis for disaster shocks.

Staff explicitly adjust their long-term baseline projections in line with the expected impact of disasters for the region times the probability of a disaster occurring in their specific country each year, subtracting this from a non-disaster projection. The projections vary given the vulnerability of a country to natural disasters—adjustments range from 0.2% of GDP up to 0.6–0.7% of GDP per year. They are largest for Vanuatu, Samoa, Solomon Islands, and Tonga. Staff reports have also discussed ways in which to help build fiscal resilience to shocks.

Further work is now looking at fiscal balances in Pacific countries using a similar cross-country panel regression methodology. This information is helpful in estimating potential fiscal buffers needed to cover budget shortfalls because of natural disasters. Staff has also looked at aid uncertainty. In the Pacific, international financial support following disasters has varied widely and unpredictably from 1.7 to 18.5% of GDP in recent years. Such uncertainty makes it exceptionally difficult for vulnerable countries to appropriately plan fiscal buffers or lines of credit needed to finance disaster recovery.

This work is aimed at helping countries better incorporate the economic impact of natural disasters into their budget and consider the types of financing needed. Like the Caribbean, insurance needs to be improved in

---

[5] The IMF's Caribbean Regional Technical Assistance Centre (CARTAC) continuously provides workshops on stress testing of the financial system, which include tests of system vulnerability to hurricanes and other plausible disaster shocks.

the Pacific at the national and regional level, and private insurance markets are largely missing for households and firms. Different international financial institutions and bilateral development partners are making efforts, but there is scope to coordinate and scale up these efforts more closely.

Source The IMF Policy Paper No. 2019/2020

5.2 THE INSURANCE MARKET AND INNOVATIVE DISASTER RISK TRANSFER MECHANISMS

According to Munich Re, one of the largest insurers of losses deriving from natural disasters, losses from natural catastrophes amounted to USD 270 billion worldwide in 2021, and roughly 55% were not insured. The insurance gap has narrowed in industrialized countries but remains substantially invariant in less developed ones. Figure 5.9 shows the trend of worldwide insured losses in the period 1970–2021.

The World Bank Group and other multilateral institutions are providing assistance to developing countries in building financial resilience

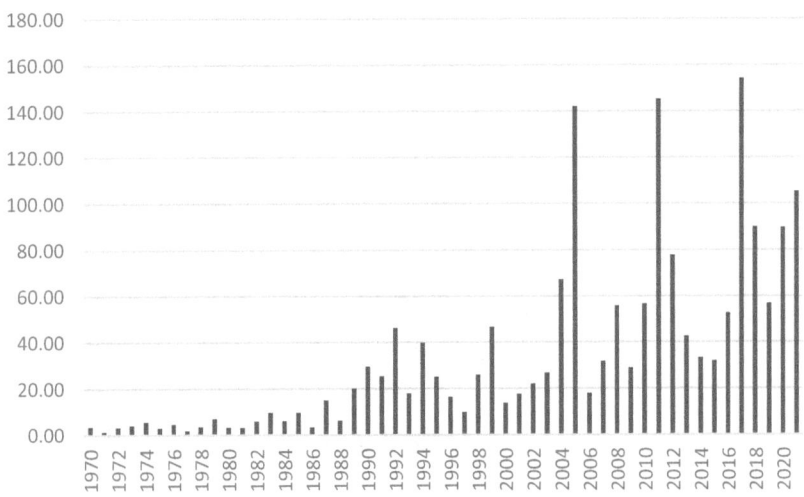

Fig. 5.9 Insured losses from natural disasters 1970–2021 (*Source* Swiss Re. 2022, retrieved from Statista)

to natural disasters through a multi-layer approach (Fig. 5.10). It combines different instruments to handle disaster risk, which could be broadly grouped into risk retention and risk transfer categories. Depending on the magnitude of the disaster, governments can decide to retain the cost through fiscal buffers, transfer risk through insurance, require contingent finance through pre-arranged credit lines, or rely on humanitarian relief from the international community.

Risk transfer mechanisms have traditionally benefited from reinsurance. It permits local insurers to retain only a tiny portion of the risk while the bulk of potential losses is transferred to international markets. This allows local economies to recover quickly from catastrophic events (Wirtz 2013). However, residents of less developed countries often refrain from buying insurance since they cannot afford it or there is a lack of social tradition of purchasing insurance. In other situations, insurance is not available due to unstable political situations.

Regarding governments, insurance coverage is still underdeveloped since substantial disaster risk insurance is perceived as expensive by sovereigns with weak fiscal positions (Fig. 5.11). The World Bank, through the GRiF, is assisting these countries in structuring viable insurance products for natural disasters, including:

- Sovereign insurance

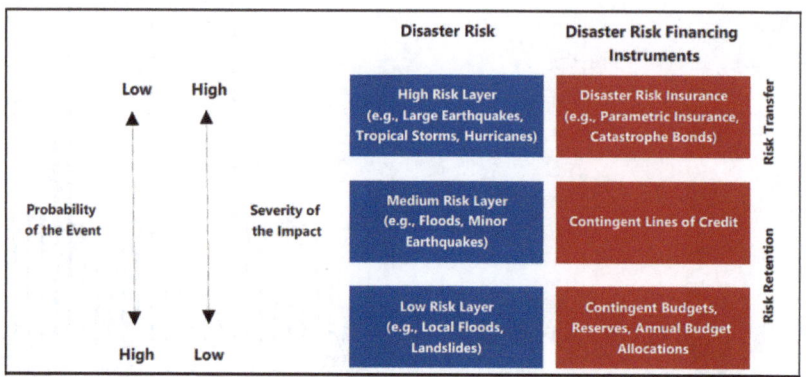

Fig. 5.10 The World Bank's multi-layer risk approach to financing disaster risk (*Source* The World Bank and the IMF)

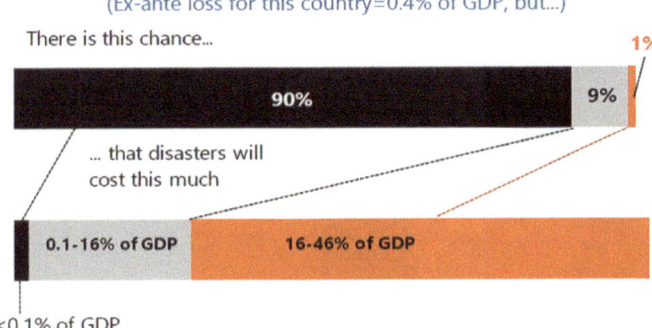

Fig. 5.11 Potential cost of tail risk natural disasters for low-income Caribbean countries (*Source* CCRIF 2021)

- Domestic insurance offered by local insurers to groups like farmers, households, or MSMEs
- Risk pools, which allow several neighboring countries to aggregate their risks of climate shocks or natural disasters.

The first risk pool in the world was the CCRIF, created in 2007. It is the first insurance instrument that develops parametric policies backed by traditional and capital markets. It provides short-term liquidity when a policy is triggered, and offers parametric insurance policies for tropical cyclones, earthquakes, or excess rainfall. The insurance products can be purchased by governments at a low cost compared to what they would pay if they had to purchase the insurance individually. Payments are triggered based on a measured parameter of the hazard event. For instance, in the case of hurricanes, this parameter is the wind speed.

Countries can purchase policies with coverage up to US$ 150 million per peril, although the limited fiscal space of countries continues to constrain their ability to do so, and many are not able to purchase optimal coverage consistent with the risk profiles of their countries, even with current discounts. Countries also receive technical assistance concerning reconstruction efforts. CCRIF contributes to the construction of flood walls and critical infrastructure and the establishment of early warning systems. Care is taken to ensure that climate and disaster risks arising from future events will be reduced.

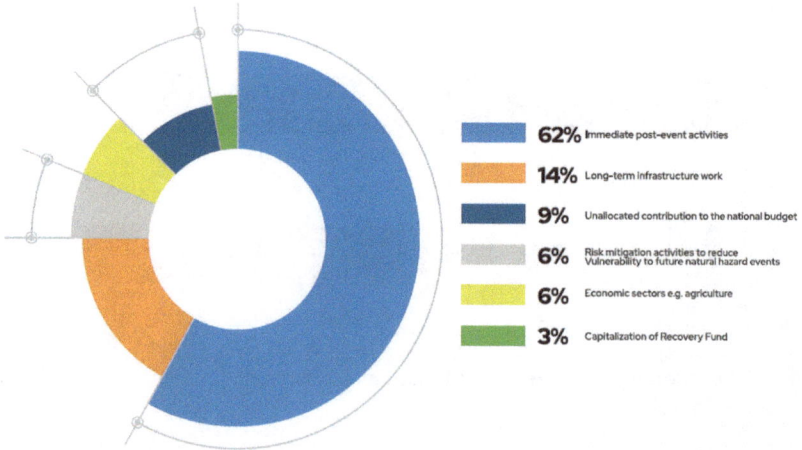

Fig. 5.12 CCRIF payouts since its inception in 2007 (*Source* CCRIF Annual Report 2021–2022)

CCRIF retains a small part of the overall risk while it transfers the most significant portion to the reinsurance market with assistance from the World Bank and other donor organizations. Since its inception in 2007, CCRIF has made 58 payouts totaling USD 260 million to 16 member governments (Fig. 5.12).

Other regions have adopted similar pooling schemes for various natural hazard risks (Table 5.2). Countries should value carefully if the higher premiums to buy insurance with high coverage outweigh the potential for solid recovery/growth post-disaster. Therefore, alternative packages could be available, which sometimes prioritize debt sustainability (lower premiums) over higher growth (lower payouts).

5.2.1 Catastrophe (CAT) Bonds and Other Insurance Solutions

CAT Bonds are securities issued by insurers and reinsurers to transfer the risk to the financial market. Investors subscribe for shares in a special purpose entity. If a particular natural disaster does not occur, he/she earns a return on top of the principal on the bond's due date, which represents a premium for the risk of the security. If the pre-defined event occurs and

Table 5.2 Regional sovereign insurance pools

	Hazards insured	Member states/territories (latest season available)	Avg. premium income/ avg. coverage
CCRIF (2007)	Earthquake Tropical cyclone (hurricanes) Excess rainfall Drought	Insured members (20): Anguilla, Antigua & Barbuda, Bahamas, Barbados, Belize, Bermuda, British Virgin Islands, Cayman Islands, Dominica, Grenada, Haiti, Jamaica, Montserrat, Nicaragua, St. Kitts & Nevis, St. Lucia, St. Vincent and the Grenadines, Trinidad & Tobago, Turks & Caicos Islands Other eligible members (15): Aruba, Costa Rica, Curacao, Dominican Republic, El Salvador, Guadeloupe, Guatemala, Guyana, Honduras, Martinique, Panama, Puerto Rico, Saint Barthelemy, Suriname, US Virgin Islands	USD 21.5 m USD 650 m
PCRAFI (2013)	Tropical cyclone Earthquake/tsunami Excess rainfall	Insured members (5): The Cook Islands, the Marshall Islands, Samoa, Tonga, Vanuatu Other eligible members (10): Fiji, Kiribati, Federated States of Micronesia, Nauru, Niue, Palau, Papua New Guinea, Solomon Islands, Timor Leste, Tuvalu	USD 2 m USD 45 m
ARC (2013)	Extreme weather (drought excess rainfall. heatwaves and tropical cyclones)	Insured members (6): Burkina Faso, Mali, Mauritania, Niger, Senegal, The Gambia Other eligible members (6): Chad, Ethiopia, Madagascar, Malawi, Kenya, Zimbabwe	USD 22 m USD 50 m
SEADRIF (2018)	Mainly flood risk	Signatories to agreement: Cambodia, Indonesia, Japan, Lao PDR, Myanmar, Singapore	TBD

Source The World Bank (2017); and IMF Policy Paper 2019/2020

triggers the CAT bond, the investor will lose all or part of the principal, which the bond issuers will use to reimburse the claims.

Jamaica provides a successful example of a CAT bond issuance in mid-2021 with the support of the World Bank. The issuance provided coverage of $185 million against hurricanes for three seasons. The World Bank also supported the development of social safety net systems and distribution channels that will be scaled up in an emergency so that proceeds from the CAT bond and other financial instruments first reach those with the greatest needs. Technical assistance provided in advance of placing the CAT bond strengthened Jamaica's disaster risk financing and insurance framework (World Bank GRiF 2021 Annual Report).[6]

State-contingent debt instruments are alternatives to insurance since they embed disaster-link clauses in debt contracts. If a country is hit by a natural disaster, interest payments or principal could be postponed for a specific period. This would avoid costly debt restructuring by these countries and reduce gross financing needs. An alternative to this instrument would be to acquire insurance against specific natural disasters to cover a specified amount of debt. If a disaster hits the country, the insurer will pay the predetermined amount for servicing the debt, reducing the financing burden of the former.

Insurance derivatives are advanced solutions that are adopted by insurers or reinsurers in developed economies. They can be similar to swaps or options. In a case of an *insurance swap*, two reinsurers can cede part of their exposure to a specific disaster risk to each other. In 2003, Swiss Re exchanged a part of its North Atlantic hurricane and European windstorm risk for Mitsui Sumitomo's Japanese typhoon exposure through a USD 100 million swap (Wirtz 2013). The insurer ceding the risk is the buyer of the option, whereas the other side (the investor) is the seller. If losses caused by the natural disaster exceed a predetermined cut-off, the investor must pay the buyer based on the terms of the contract.

Weather index insurance is a type of derivative purchased by the insured entity. It can decide the optimal coverage, and the payment received in case of disaster is not linked to the actual losses but rather to the coverage. This avoids the claim validation process, which is very time-consuming.

Microinsurance offers protection to low-income people against specific perils, such as weather hazards, in exchange for premiums that are proportionate to their cost of living and the cost of the risk involved (International Association of Insurance Supervisors). These products target communities with little or absent insurance culture. Premiums are

---

[6] The premiums to the insurers were paid by the World Bank. Jamaica is expected to pay for the renewal itself, since the CAT bond should give time to the country to reduce its debt.

calculated using group pricing, and payments need not be regular. Often microinsurance is sold in bundle packages with microloans from microfinance institutions or mobile phone airtime. Therefore, microinsurance does not target the bottom of the pyramid (the poorest of the poor) but those leaving poverty, which could suffer huge losses from specific perils. Box 5.4 describes the case of Esfuerzo Seguro, a microinsurance scheme provided by the Microinsurance Catastrophe Risk Organization (MiCRO) in Guatemala. MiCRO was established in 2011 by Mercy Corps and Fonkoze as a specialized reinsurance company incorporated in Barbados after the 2010 Haiti earthquake.

Box 5.4: Esfuerzo Seguro

Description:
Esfuerzo Seguro is a holistic risk management solution offered by Aseguradora Rural to agro and non-agro clients of Banrural in Guatemala when contracting a productive credit line. MiCRO's approach has two major components: (1) an index-based microinsurance product and (2) a value-added program that includes tools to raise awareness about disaster risk preparedness and a financial education program to empower consumers.

It is the **first index-based insurance** product in Central America, protecting vulnerable and low-income segments of the population against the business interruption caused by earthquakes, droughts, and severe rainfall.

Benefits and covered perils:
Esfuerzo Seguro covers the business interruption of productive activity against excessive rainfall, severe drought, and earthquakes. Payouts vary based on the level of deviation of the corresponding index from historical averages.

MiCRO's calculation platform ("MiCAPP") continuously monitors and extracts data from predetermined scientific sources (e.g., NASA, USGS), unpacks the scientific data files and calculates them using complex algorithms to match the readings against predetermined levels, and issues a loss report when a triggering event is detected, which is accepted by all stakeholders as the undisputable base for payouts.

Premium: 5% of the value of the initial credit (5.6% with IVA)
Date of launch: November 2016
Pilot phase: November 2016–December 2017
(2633 clients during the pilot phase)

Number of policies: More than 4000 clients by April 2018

Project background: Esfuerzo Seguro was designed and implemented in the context of the Central American Disaster Microinsurance Expansion (CADME), a program executed by MiCRO with the support of the Swiss Development Agency (SDC), the Multilateral Investment Fund (MIF) managed by the Inter-American Development Bank (IADB), the Australian Aid, Swiss Re, Mercy Corps, and KfW through its Climate Adaptation Platform (CAP), a genuinely public–private partnership.

5.2.2   Government-Sponsored Pools for Natural Disasters to Overcome Problems of Market Failure

Several advanced economies have established government-sponsored insurance pools against specific perils as an alternative to individual insurance purchases, which is often extremely costly and escalates rapidly. Some examples are the National Flood Insurance Program (NFIP in the U.S.), Florida Hurricane Catastrophe Fund, California Earthquake Authority, and New Zealand's Earthquake Commission. In emerging markets, the Turkish Catastrophe Insurance Pool, established in 2000 with World Bank assistance, manages the compulsory earthquake insurance in the country, limiting the fiscal contingent liability. The World Bank, through the GRiF program, is assisting several developing countries in establishing local insurance pools against certain natural disasters, similar to what was done for Turkey in 2000.

In the following the NFIP will be described shortly. It was established in 1968 by the U.S. Congress and has two purposes, to share the risk of flood losses through flood insurance and to reduce flood damages by restricting floodplain development. The program enables property owners in participating communities to purchase insurance protection administered by the government, against losses from flooding. It requires flood insurance for all loans or lines of credit secured by existing buildings, manufactured homes, or buildings under construction located in the Special Flood Hazard Area in a community that participates in the NFIP. The U.S. Congress limits the availability of National Flood Insurance to communities that adopt adequate land use and control measures with effective enforcement provisions to reduce flood damages by restricting development in areas exposed to flooding.

As of August 2017, the program insured about 5 million homes (down from about 5.5 million in April 2010), most of which are in Texas and Florida. The cost of the insurance program was fully covered by its premiums until the end of 2004. However, following Hurricane Katrina and Sandy, it accumulated $25 billion of debt by August 2017 (Witkowski and Scism 2017). In October 2017, the U.S. Congress canceled $16 billion of NFIP debt, making it possible for the program to pay claims. The NFIP owes $20.525 billion to the U.S. Treasury as of December 2020.

The program has attracted increased criticism because of problems of adverse selection. Specific provisions within the NFIP increase the likelihood that flood-prone properties will be occupied by the people least likely to be in a position to recover from flood disasters, which further increases the demand for aid. According to critics of the program, the government's subsidized insurance plan "encouraged building, and rebuilding, in vulnerable coastal areas and floodplains." Stephen Ellis of the group Taxpayers for Common Sense points to "properties that flooded 17 or 18 times that were still covered under the federal insurance program" without increasing premiums. Another criticism is that FEMA only administers some policies but outsources many policies to private insurance companies. FEMA pays private insurance companies to offset their costs when a disaster occurs. However, there needs to be more oversight and rules regarding how the money should be distributed. Consequently, private insurers use FEMA payments to hire attorneys that fight policyholders in court. One law firm is estimated to have received US$29M from FEMA payments to fight Hurricane Sandy claims (CBS News, 30 April 2018).

5.3 The Role of the Government in Natural Disaster Resilience and Recovery

Various sections of this study have acknowledged that building resilience to weather hazards can only be successful with government action at the central or local level. In this section, the focus will be on the most important trends observed worldwide regarding the role of public finance in disaster management. It must be stressed that the government's role goes beyond providing aid to communities hit by disasters. Instead, governments need to build an adaptation and risk management culture in vulnerable countries, given the increased frequency and magnitude of weather hazards across the globe.

It is essential first to clarify that disaster risk management strategies are usually categorized into Disaster Risk Reduction (DRR) and Climate Change Adaptation (CCA) groups. DRR aims to prevent new and reduce existing disaster risk and manage residual risk. CCA regards the process of adaptation to climate change. Climate change harms the communities, but potential benefits could also materialize (Choi et al. 2023). Figure 5.13 shows that the two approaches overlap, but there are also differences, mainly because DRR deals with all types of disasters, whereas CCA focuses on climate-related ones. Moreover, CCA is also related to chronical risks such as mean temperature increase or sea level rise.

Although, recently, calls for integrating these two strategies are increasing, many countries still treat them separately regarding public finance. Various organizations such as the UNFCCC, the IPCC, the

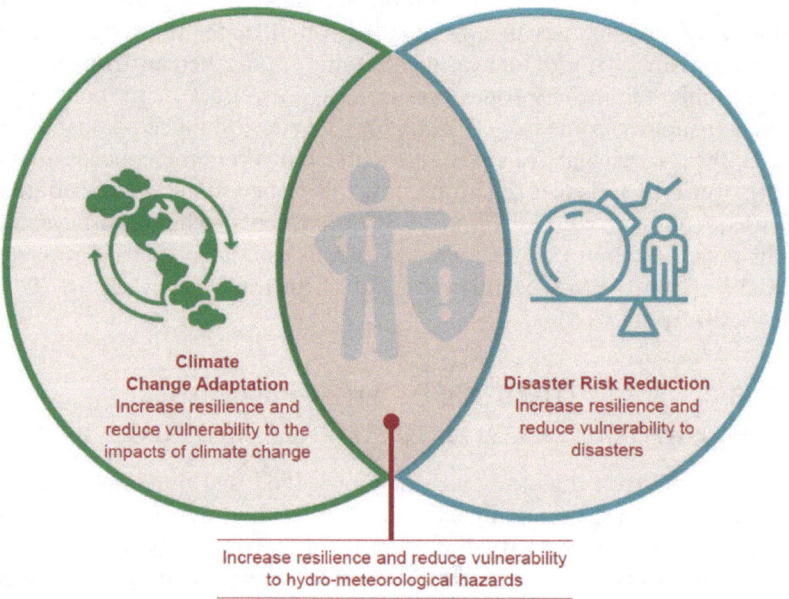

Fig. 5.13 Overlap and differences between climate change adaptation and disaster risk reduction (*Source* Choi et al. 2023)

UNDRR, and OECD advocate comprehensive disaster and climate risk management.

Previously it was underlined that developing countries face a widening gap between their adaption needs and available funding for climate disaster resilience. What is also worrying is that funding usually goes to mitigation rather than adaptation projects which need clear targets to achieve by national governments or the international community. Hence, most funds for DRR projects originate from domestic public expenditure. However, countries have high variability in the amount spent and how resources are allocated. Three-quarters of these investments do not directly target DRR. Therefore they are indirect expenditures.

Many countries have adopted the Sendai Framework for Disaster Risk Reduction (Sect. 5.1). However, the targets set by the Framework are not quantified and are not legally binding.[7]

To date, many countries have introduced climate strategies or policies. The Grantham Research Institute of Climate Change and the Environment found 2507 climate-related laws and policies at the beginning of 2022. It is worth noting that the number of laws and policies introduced in each country does not indicate the country's effectiveness in addressing climate change. This is instead determined by the scope of the laws and policies in a country and the ability to implement them (Eskander et al. 2020). A joint climate change and disaster risk strategy have been adopted by 13 countries, while other countries recognize that CCA and DRR are part of other strategies and plans (Table 5.3).

Regarding the different continents, various countries consider the importance of synergies between CCA and DRM measures. European countries already include climate change in disaster risk assessments. In Africa, several countries have implemented disaster and policy review tools for different time periods. In Asia and the Pacific region, this exercise has been done mainly by island states (Kiribati, Marshall Islands, Micronesia, Nauru, Palau, Papua New Guinea, Solomon Islands, Tonga, and Vanuatu). In contrast, in Latin America, there are the best practices of Colombia, Ecuador, and Peru.

Countries are starting to tag their public expenditures related to disasters and climate change. This is done to raise awareness, increase

---

[7] Many countries have adopted the Sendai Framework for Disaster Risk Reduction (Sect. 5.1). However, the targets set by the Framework are not quantified and are not legally binding.

Table 5.3 Joint climate change and DRR strategies and plans

Country	Strategy or plan
Cambodia	Climate Change Strategic Plan for Disaster Management Sector 2014–2018
Cook Islands	Joint National Action Plan for Disaster Risk Management Climate Change Adaptation (2016–2020); Climate and Disaster Compatible Development Policy 2013–2016 (Kaveinga Tapapa)
Egypt	Egypt's National Strategy for Adaptation to Climate Change and Disaster Risk Management
Kiribati	Kiribati Joint Implementation Plans for Climate Change and Disaster Risk Management 2014–2023 and 2019–2028
Maldives	Strategic National Action Plan for Disaster Risk Reduction and Climate Change Adaptation 2010–2020
Marshall Islands	Joint National Action Plan for Climate Change Adaptation and Disaster Risk Management 2014–2018
Micronesia	Nationwide Integrated Disaster Risk Management and Climate Change Policy
Nauru	Framework for Climate Change Adaptation and Disaster Risk Reduction
Nepal	Priority Framework for Action: Climate Change Adaptation and Disaster Risk Management in Agriculture 2011–2020
Niue	Joint National Action Plan for Disaster Risk Management and Climate Change
Tonga	Joint National Action Plan on Climate Change Adaptation and Disaster Risk Management 2010–2015
Tuvalu	National Strategic Action Plan for Climate Change Adaptation and Disaster Risk Management 2012–2016
Vanuatu	The Vanuatu Climate Change and Disaster Risk Reduction Policy 2016–2030; National Policy on Climate Change and Disaster-Induced Displacement

Source Choi et al. (2023)

transparency and accountability, assess trends, and identify gaps, and available resources. Several countries use online portals or citizens' budgets to disseminate climate-related budget allocation and expenditure information to the broader public (Uganda, Philippines, Ecuador, Colombia, Ghana, etc.). Figure 5.14 evidences Colombia's budget allocation toward DRM in the period 2011–2019.

Countries report capacity constraints on climate and DRR concepts and how to tag CCA and DRR during budget preparation. Maintaining continuous capacity is resource intensive and can compromise the reforming process. Training staff on green budgeting frameworks

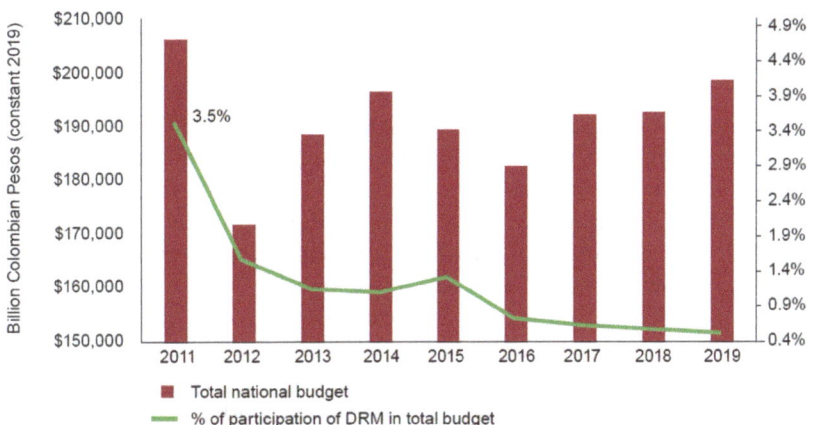

Fig. 5.14 Colombia's budget allocation towards DRM (2011–2019) (*Source* Choi et al. 2023)

and practices is one of the critical issues in the European Union, but not only. The centralization/decentralization level of the country impacts the implementation of CCA/DRR into public financial management and budgeting. Highly centralized administrations could face a significant burden in monitoring and reporting activities.

Various initiatives have been undertaken to overcome capacity constraints. Several countries rely on technical support from regional organizations and development partners (Armenia, Azerbaijan, Georgia, Kyrgyzstan, Kazakhstan, and Serbia for example). The risk of this approach is that external advisors need to comprehend the national context better. Otherwise, the top-down initiatives become a sort of academic exercise. Some countries have ensured that capacity gaps can be addressed by explicitly incorporating capacity-building provisions within their tagging and tracking frameworks to address capacity gaps (the Philippines, for example).

The experience matured by several countries has shown that a solid legal framework and political support are necessary conditions for embedding CCA/DRR in public fiscal frameworks, tracking expenditures, and monitoring the success of the projects that are being implemented. The problem of political support derives from the many priorities that governments of less developed countries have to bring forward, such as social protection, welfare, youth, etc. Moreover, it is necessary to coordinate various stakeholders at the central and local levels (for example, Ministries of Finance, Ministries of Environment, civil protection agencies, local government units, etc.).

As mentioned before, the challenging and most important task for governments is to establish a risk reduction culture in the country that goes beyond incorporating DRR in the budgeting framework. Governments should address equality of income, equity, and fairness issues by assisting residents and small businesses financially so they can afford to invest in DRR. At the most basic level, governments need to ensure regulations are in place to prevent, reduce or ensure the resilience of construction in unsafe locations, such as floodplains, areas subject to sea level rise, or areas at extremely high risk of fire or other hazards.

Governments also need to understand the climate projections for their jurisdictions better. They should work with experts to update design standards to ensure resilient infrastructure design, particularly against increased temperatures, higher-intensity rainfall, and drought impacts. In parallel, it is essential to assess the risks to current critical infrastructure under a range of future scenarios likely to occur within their lifetime. Implementing these cost-effective protection measures can help reduce the need for costly humanitarian assessments, thus saving money and suffering.

Governments should improve the targeting of humanitarian assistance to the most vulnerable through grant-based or longer-term assistance through loans. This can help incentivize future risk reduction. Encouraging the private sector to review the resilience and sustainability of systemic risks of their operations can send necessary signals to encourage preparedness. Also, encouraging insurers to protect against losses from disasters by supplying reinsurance coverage against catastrophic losses for those who take preventive measures can help ensure safety nets are in place (van den Bergh and Botzen 2020).

The public sector is also crucial in creating a new "social contract" to incentivize investment in disaster resilience. It can help specify the responsibilities and liabilities of national governments, financing bodies, and the private sector to manage the negative externalities arising from disaster risks. National governments and regulators must define sustainable, disaster-resilient investments and encode risk metrics to change investor behavior and raise awareness of disaster risks.

Box 5.5: Green Bonds for Climate Resilience
Since the first labeled green bond in 2007 by the European Investment Bank, $1.5 trillion of labeled green bonds has been issued worldwide from a diverse range of issuers, including sovereigns, municipalities, national development banks, financial institutions, and corporates. About 16.4% (1265) of green bonds (7725 deals) have included activities related to adaptation and resilience, mainly in the water and water-related sectors. Of these, 79% were issued by developed markets, 15% from supranational institutions, and only 6% from emerging markets (Qadir et al. 2021).

Recent examples include:

The city of Malmö in Sweden, one of the earliest municipal green bond issuers, used two issuances to raise funds for climate change adaptation and resilience measures for sustainable management of water, wastewater, land, and natural resources.

The Asian Development Bank issued a bond in 2019 that prominently featured adaptation and resilience activities. Investments include the Mongolian Ulaanbaatar Green Affordable Housing and Resilient Urban Renewal Sector Project, which is building 10,000 energy-efficient and low-carbon housing units as part of 20 new eco-districts with resilient infrastructure like roads, water, sewerage, heating pipes, and greenhouses for urban farming.

Grupo Rotoplas, a corporate entity in Mexico, issued a $523 million green bond in 2017 that included resilience finance for innovative water solutions in markets where clean water is scarce due to droughts, water pollution, and unreliable water infrastructure.

The annual climate change adaptation costs for developing countries are estimated to be in the range of $140 billion to $300 billion per year by 2030 and between $280 billion and $500 billion per year by 2050 if global warming is limited to 2 °C above pre-industrial levels (UNEP 2016).

In the face of these needs, adaptation finance flows still need to be more robust. Total tracked public and private investment in climate adaptation in 2018 was $30 billion worldwide (Buchner et al. 2019). Public finance will be insufficient to meet adaptation financing needs, particularly in developing countries. While there is limited data on private investment flows, securing private investment for adaptation remains challenging. Climate resilience bonds (Box 5.5) could help increase investment in adaptation and accelerate a resilient, sustainable climate transition (Qadir et al. 2021).

REFERENCES

Adshead, D., Fuldauer, L.I., Thacker, S., Román García, S., Vital, S., Felix, F., Roberts, C., Wells, H., Edwin, G., Providence, A., Hall, J.W., 2020. Saint Lucia: National Infrastructure Assessment. Copenhagen: United Nations Office for Project Services.

AfDB, & Asian Development Bank, & EBRD, & EIB, & Inter-American Development Bank, & World Bank, & IsDB, 2022. Climate finances from multilateral development banks globally in 2021, by activity (in million U.S. dollars).

Buchner, B., Clark, A., Falconer, A., Macquarie, R., Meattle, C., Tolentino, R., Wetherbee, C., 2019. Global Landscape of Climate Finance. London: Climate Policy Initiative. Accessible at: www.climatepolicyinitiative.org/wp-content/uploads/2019/11/2019-Global-Landscape-of-Climate-Finance.pdf

CBS News., 2018. Federal program meant to help flood victims spends millions fighting claims. Accessible at: https://www.cbsnews.com/news/national-flood-insurance-program-meant-to-help-victims-spends-millions-fighting-claims/

CCRIF., 2021. Annual Report 2021–2022. Accessible at: https://www.ccrif.org/publications/annual-report/ccrif-spc-annual-report-2021-2022

Choi, S., Weingärtner, L., Gaile, B., Cardenas, D., Wickramasinghe, K., Nicholson, K., Broermann, S., Tchané, Y.B., Steele, P., 2023. Tracking the money for climate adaptation and disaster risk reduction. UNDRR, Accessible at: https://www.undrr.org/publication/tracking-money-climate-adaptation-and-disaster-risk-reduction.

EIB, 2021. Joint report on multilateral development banks' climate finance. Accessible at: https://www.ebrd.com/news/2022/2021-sees-record-joint-mdb-climate-finance-.html

Eskander, S., Fankhauser, S., Setzer, J., 2020. Global lessons from climate change legislation and litigation. Environmental and Energy Policy and the Economy, 2, pp. 44–82.

Fuldauer, L., Thacker, S., Hall, J., 2021. Informing national adaptation for sustainable development through spatial systems modelling. Global Environmental Change, 71, p. 102396.

GFDRR., 2022. Annual Report. Accessible at: https://www.gfdrr.org/en/publication/gfdrr-annual-report-2022

GRiF., 2021. Annual Report. Accessible at: https://www.globalriskfinancing.org/publication/global-risk-financing-facility-grif-annual-report-2021

IMF., 2019. Building resilience in developing countries vulnerable to large natural disasters. IMF Policy Paper, 2019/2020.

Lee, D., Zhang, H., Nguyen, C., 2018. The economic impact of natural disasters in Pacific island countries: Adaptation and preparedness. IMF Working Paper No. 18/108, Washington.

Munich Re Group., 1999. Topics 2000: Natural Catastrophes. Munich, Germany, pp. 70–112.

Pant, R., Hall, J.W., Koks, E.E., Paltan, H., Hu, X., Zorn, C., Russell, T., 2022. From local to global scales—quantifying climate risks and adaptation opportunities for networked infrastructure systems. GAR2022 Contributing Paper. Geneva: United Nations Office for Disaster Risk Reduction.

Qadir, U., Pillay, K., Creed, A., Boulle, B., Tukiainen, K., Lala, O., Tapia, M., Molloy, D., Rae, J.J., 2021. Green Bonds for Climate Resilience: State of Play and Roadmap to Scale. Rotterdam: Global Center on Adaptation.

Swiss Re, 2022. Insured losses caused by natural disasters worldwide from 1970 to 2021 (in billion U.S. dollars).

The World Bank, 2017. Sovereign catastrophe risk pools: World Bank technical contribution to the G20. Available on the internet: https://openknowledge. worldbank.org/handle/10986/28311.

UNDRR., 2022. Global Assessment Report. Accessible at: https://www.undrr. org/gar2022-our-world-risk-gar.

UNEP., 2016. The Adaptation Gap Report. Accessible at: https://www.unep. org/resources/adaptation-gap-report-2016.

UNEP., 2022. The Adaptation GAP Report. Accessible at: https://www.unep. org/resources/adaptation-gap-report-2022.

van den Bergh, J., Botzen, W., 2020. Low-carbon transition is improbable without carbon pricing. Proceedings of the National Academy of Sciences, 117, pp. 23219–23220.

Wirtz, A., 2013. Natural disasters and the insurance industry. In Guha-Sapir, D., Santos, I., (Eds.) The Economic Impact of Natural Disasters. Oxford Academic.

Witkowski, R., Scism, L., 2017. Hurricane Harvey threatens largest flood insurer: The Government. Wall Street Journal.

Zapata-Marti, R., 2013. Natural disaster mitigation policies. In Guha-Sapir, D., Santos, I., (Eds.) The Economic Impact of Natural Disasters. Oxford Academic.

CHAPTER 6

Conclusions

Abstract Severe weather hazards pose significant challenges to the economies of many countries, and their impact will be even more significant in the future. Research on the economic consequences of natural disasters suggests that they impact short-term growth, but opinions diverge on long-term outcomes. At the same time, this literature finds that various country-specific factors matter in facilitating post-disaster recovery. While this body of research is vibrant, it has yet to deal adequately with the consequences of severe weather events on banking institutions and potential spillover effects on the real economy. Natural disasters could pose severe threats to banks depending on their area of operations, especially to small institutions in emerging countries. Governments of these countries need to work closely with international cooperation agencies, insurance companies, and the scientific community to multiply funding toward climate change adaptation projects and establish a climate risk reduction culture in their societies.

Keywords Banking and the real economy · Weather hazards · Financial sector stability · Central banks

Natural disasters pose significant challenges to many countries, and their impact will be even more significant in the future. Severe weather hazards generate social and economic adverse effects, especially for less developed economies. The short and long-run economic effects of disasters have attracted the interest of researchers and policymakers. There is consistent and diversified literature on the disaster–growth nexus. The empirical outcomes suggest that disasters impact short-term growth, but opinions diverge on long-term consequences. Short-term losses are likely followed by an expansion when capital is upgraded with more productive versions, and this technological change increases productivity. Alternatively, the economy returns to its long-term path, although there could be a temporary push from an inflow of FDI, aid, insurance claims, and reconstruction investments. However, another scenario assumes that some countries hit by major and frequent disasters cannot return to their previous long-term track because of financial constraints, which force firms and households to underinvest.

At the same time, this literature finds that various country-specific factors matter in facilitating post-disaster recovery. These can be related to the pre-disaster level of development of the country, its financial system, the quality of its institutions, country size, geographic location, financial liberalization, etc.

While this body of research is vibrant, it has not dealt adequately with the consequences of severe weather events on banking institutions and potential spillover effects on the real economy. In Chapters 2 and 3 of the book, we first aimed to show that banking institutions face higher risks from natural disasters but also play a significant role in addressing credit needs for reconstruction and post-disaster recovery.

Climate change is perceived to have a considerable impact on the economy and also on the stability of the financial sector. Banking regulators and policymakers have started to discuss the role of the banks in fostering a greener economy and how banks might be affected by environmental hazards, which will be even more severe and catastrophic in the future. Central banks are assessing the potential impact of climate scenarios on bank stability. In these scenarios, acute physical risks such as natural disasters could pose serious threats to banks depending on their area of operations. They could hamper firm revenues, operational costs, and capital expenditures by increasing investments to cope with climate change. These outcomes could spill over to the banking sector in multiple ways, such as through the depreciation of collateral, misvaluation of securities of the trading book, losses from branch closures, or liquidity pressure from the redemption of deposits. In the medium and long term, the average quality of banks' customer base could deteriorate due to worsening economic conditions. At the same time, acute events such as hurricanes or floods generate income for banks due to increased credit demand for reconstruction needs.

Central banks are trying to understand how these trends might evolve through the adoption of various climate models. However, there are still various complexities in the link between climate change and banking activity since models need to be more sophisticated, and many countries need more climate data about their territories. Most physical risk analyses focus on direct physical damage to properties and infrastructure, with little reference to other variables that affect firms' operating environment, such as supply chain linkages.

What emerges from the academic literature is that natural disasters increase bank credit and liquidity risk in emerging markets and for relatively small regional banks operating in developed countries such as the U.S. or Germany. Generally, banks' probabilities of default after natural disasters tend to increase, together with non-performing loans and foreclosure ratios. It seems that banks need to store liquidity to handle the pressure originating from a decrease in deposits, and they are starting to price climate risk in their products through a tightening of credit standards and more stringent loan terms for potential exposures to certain types of physical risks (hurricanes or sea level rise for instance).

There is also a body of literature that looks at banks' role in promoting economic recovery after disaster strikes through the bank lending channel. The outcomes of this research confirm that bank credit is vital for reconstruction needs since regions, where credit is not interrupted recover faster from the disaster. Even in this case, this mechanism is spurred by other bank or country factors. Banks do not reduce credit when well-capitalized and enjoy a reasonable degree of market power in certain areas. This is usually observed for small regional banks, which concentrate lending in a restricted geographical area, or large banks that maintain a branch presence in that area. At the same time, in advanced countries such as the U.S., banks are supported by a well-functioning securitization market, which allows them to transfer climate risks to financial markets.

The book's last chapter looks at other actors essential in alleviating disaster damage and increasing countries' resilience to natural disasters.

These are multilateral development banks, other international organizations, the insurance sector, and governments. This chapter focuses mainly on emerging markets that cannot cope with disasters' consequences compared to more advanced economies. In recent years, multilateral banks such as the World Bank have multiplied their funding toward climate finance, although the focus has been primarily on mitigation rather than adaptation projects. At the same time, this institution has launched several programs, funds, or facilities to support less developed countries to find sustainable solutions from a financial point of view to the consequences of natural disasters. These alternatives are jointly created with partners from developed countries, insurance companies, and other financial institutions.

Governments in almost every country have begun to integrate climate policies into their legislations and public fiscal frameworks. The International Monetary Fund assists them in assessing the best ways to adapt and mitigate climate change. Depending on the magnitude of disasters, governments can retain the cost through fiscal buffers, transfer the risk through insurance, or rely on humanitarian relief from the international community.

Governments' most complex and essential task is establishing a climate risk reduction culture in their countries. This can be achieved by taking several steps, such as enforcing regulations about the resilience of construction in disaster-prone areas, understanding their countries' exposure to several climate risks by working closely with the scientific community, ensuring that financial assistance goes to all segments of the population so that even the most vulnerable invest in disaster risk reduction, taking measures to improve humanitarian assistance and reconstruction loans or grants after a disaster, encouraging insurers to provide protection from disasters to those more exposed by supplying reinsurance coverage against catastrophic losses.

INDEX